REVISE SALTERS-NUFFIELD AS/A LEVEL
Biology A

REVISION WORKBOOK

Series Consultant: Harry Smith

Author: Gary Skinner and Ann Skinner

Reviewer: David Barrett

A note from the publisher

In order to ensure that this resource offers high-quality support for the associated Pearson qualification, it has been through a review process by the awarding body. This process confirms that this resource fully covers the teaching and learning content of the specification or part of a specification at which it is aimed. It also confirms that it demonstrates an appropriate balance between the development of subject skills, knowledge and understanding, in addition to preparation for assessment.

Endorsement does not cover any guidance on assessment activities or processes (e.g. practice questions or advice on how to answer assessment questions), included in the resource nor does it prescribe any particular approach to the teaching or delivery of a related course.

While the publishers have made every attempt to ensure that advice on the qualification and its assessment

is accurate, the official specification and associated assessment guidance materials are the only authoritative source of information and should always be referred to for definitive guidance.

Pearson examiners have not contributed to any sections in this resource relevant to examination papers for which they have responsibility.

Examiners will not use endorsed resources as a source of material for any assessment set by Pearson.

Endorsement of a resource does not mean that the resource is required to achieve this Pearson qualification, nor does it mean that it is the only suitable material available to support the qualification, and any resource lists produced by the awarding body shall include this and other appropriate resources.

> **For the full range of Pearson revision titles across KS2, KS3, GCSE, AS/A Level and BTEC visit:**
> www.pearsonschools.co.uk/revise

ALWAYS LEARNING

PEARSON

Contents

A small bit of small print

Edexcel publishes Sample Assessment Material on its website. This is the official content and this book should be used in conjunction with it. The questions in Now Try This have been written to help you practise every topic in the book. Remember: the real exam questions may not look like this.

1-to-1 page match with the SNAB Biology A Revision Guide ISBN 9781447992714

Why is transport needed?

Guided   **1** Complete this table about some important features of water.

Feature	Explanation
liquid at room temperature	Water molecules are joined to each other by ....
polar solvent	....
high specific heat capacity The amount of energy needed to raise the temperature of water is high.	....

(5 marks)

Guided   **2** Imagine a cube with volume 1 cm³.

(a) Calculate its total surface area.

....

....   **(1 mark)**

Now imagine the same volume 'squashed' so it forms a euboid with dimensions 4 cm × 4 cm × 0.0625 cm.

(b) Calculate the new total surface area.

....

....   **(1 mark)**

(c) Which shape would make the better exchange surface? Explain your answer.

....

....

....

....   **(3 marks)**

Blood vessels

1 (a) Draw a labelled diagram to show the structure of an artery.

> Keep diagrams simple and clear and make sure label lines touch the labelled object.

(4 marks)

(b) Explain how the artery is adapted to its function.

> When talking about **structure** and **function**, you need to say it is like **this** because it does **this**.

....

....

....

....

(4 marks)

2 The photograph on the left shows a cross-section of two blood vessels, labelled **A** and **B**. The diagram on the right shows a human heart with the positions of two of the blood vessels numbered **1** and **2**. Match each blood vessel with its position in the heart.

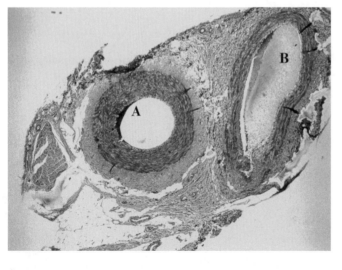
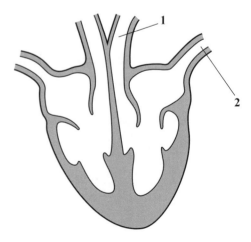

A ....

B ....   **(2 marks)**

The heart

1 *Describe the structure of the mammalian heart.

....

....

....

....

....

....

....

....

....   **(6 marks)**

> In starred questions (*), structure your answer logically showing how the points you make are related to or follow on from each other. You need to select and apply relevant knowledge of biological facts or concepts to support the argument.

> There are many points to make here. It might be best to list each point and then put the list into prose to prevent rambling.

2 (a) State the function of valves in the circulatory system.

....   **(1 mark)**

   (b) Explain why valves are found where they are in the circulatory system.

....

....

....

....

....

....   **(4 marks)**

> This is an 'Explain' question, so it is not enough to state your point – you must also say why you have come to that conclusion.

3 Explain why the wall of the left ventricle is thicker than that of the right ventricle.

....

....

....   **(2 marks)**

The cardiac cycle

1 Describe the stages in the cardiac cycle.

The cardiac cycle consists of three stages: atrial systole, ventricular

systole and ....

During atrial systole, the .... contract and the

.... are relaxed.

The .... valves are open.

During ventricular systole, the .... open as oxygenated

blood is forced out of the heart through the aorta to the body and

through the pulmonary .... to the lungs.   **(3 marks)**

2 The graphs show pressure changes and an ECG during one cardiac cycle of the left side of the heart.

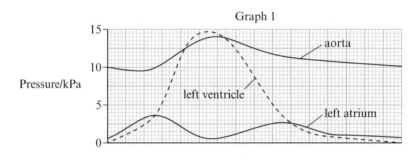

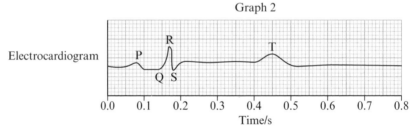

Maths skills

(a) Calculate the heart rate.

Heart rate = ...   **(2 marks)**

(b) After how many seconds does each of the following happen?

(i) atrioventricular valve closes ....

(ii) aortic (semilunar) valve closes ....

(iii) atrioventricular valve opens ....   **(3 marks)**

(c) Describe what is happening during QRS and T phases of the ECG.

....

....   **(2 marks)**

Clots and atherosclerosis

Guided

1 *Describe the sequence of events in the clotting process.

A blood clot may form when ....

....

....

A series of chemical changes occur in the ....

....

....

....

These strands form ....

....

.... **(6 marks)**

> In starred questions (*), structure your answer logically showing how the points you make are related to or follow on from each other. You need to select and apply relevant knowledge of biological facts or concepts to support the argument.

2 Explain how atherosclerosis develops.

....

....

....

....

....

.... **(4 marks)**

> This kind of answer could be set out as a flowchart of events or bullet points – both are acceptable.

3 Explain how atherosclerosis can cause the chest pain associated with an attack of angina.

....

....

....

.... **(3 marks)**

> You need to think about the consequences of a lack of proper circulation in terms of what the blood delivers to tissues.

Risk

1 State **two** dietary factors that increase the risk of CVD (cardiovascular disease).

> Note the word 'dietary'.

.... **(1 mark)**

2 State what is meant by the term risk factor.

.... **(1 mark)**

3 Complete this diagram by giving **three** other ways in which the risk of CVD may be reduced.

```
┌──────────────┐                    ┌──────────────────────┐
│              │                    │  lowering blood      │
│              │                    │  cholesterol levels  │
└──────────────┘                    └──────────────────────┘
         ↘                          ↙
              ┌──────────────────┐
              │ reducing the risk│ ←──  ┌──────────────────┐
              │    of CVD        │      │                  │
              └──────────────────┘      └──────────────────┘
         ↗                          ↖
┌──────────────────────┐     ┌──────────────────────┐
│ lowering blood pressure│    │                      │
└──────────────────────┘     └──────────────────────┘
```

(3 marks)

Maths skills

4 It has been shown that for men aged 40 to 50 years old, each rise of 10 units in their systolic blood pressure increases the risk of heart disease by 20%. Calculate the increased risk of heart disease in a 40-year-old man with systolic blood pressure 50 units higher than the average for his age.

Answer: .... **(1 mark)**

Guided

5 *Explain why the combination of a high-fat diet and low activity levels may lead to CVD.

> In starred questions (*), structure your answer logically showing how the points you make are related to or follow on from each other. You need to select and apply relevant knowledge of biological facts or concepts to support the argument.

Due to an energy imbalance the person becomes obese

....

....

....

....

.... **(6 marks)**

Correlation and causation

1 Data on the cholesterol levels and blood pressure for different adult populations in America were collected. The mean cholesterol level and the percentage of each population with high blood pressure were calculated. The results are shown in the table.

Adult population (ethnic groups)	Mean cholesterol level/ mg dm⁻³	Percentage of population with high blood pressure (%)
A	214	42
B	216	28
C	215	30

(a) Describe the relationships between blood cholesterol level, ethnic group and percentage of population with high blood pressure.

> Note that the word 'relationships' is plural so you need to look at more than one.

....

....

....

....

.... **(3 marks)**

(b) A student concluded from these data that high blood pressure has a genetic cause. Explain why this conclusion may **not** be valid.

....

....

....

....

....

.... **(4 marks)**

2 Distinguish between the terms causation and correlation.

....

....

....

....

.... **(2 marks)**

Studying the risks to health

1   Compare and contrast cohort studies with case-control studies.

....

....

....

....   **(3 marks)**

> You must discuss both similarities and differences in 'compare and contrast' questions and look at both angles.

2   Describe the key features of a good study used to determine health risk factors.

....

....

....

....

....   **(4 marks)**

> You should write about four features.

Guided

3   *Data were collected about the rate of CHD (coronary heart disease, an aspect of CVD) in France and the UK. The graphs show these data and some information about lifestyle factors known to be involved in CHD. CHD death rates in France are lower than might be expected from the lifestyle data. This is called the 'French Paradox'. Analyse the data to explain how the idea of a 'French Paradox' is supported.

> In starred questions (*), structure your answer logically showing how the points you make are related to or follow on from each other. You need to select and apply relevant knowledge of biological facts or concepts to support the argument.

The UK has a higher death rate from CHD than France by about seven

times ....

....

....

....

....   **(5 marks)**

Energy budgets

Maths skills

1 One way of defining obesity is to use the body mass index (BMI).
Calculate the BMI for an individual who weighs 76 kg and whose height
is 1.70 m and state if this individual would be described as obese.

> BMI = body mass in
> kg ÷ (height in m)²

BMI = ...   **(2 marks)**

2 The chart shows the effect of having a high BMI on the risk of death due to
CHD compared with a control group with a BMI of 20. **Analyse** this information
to describe the trend shown.

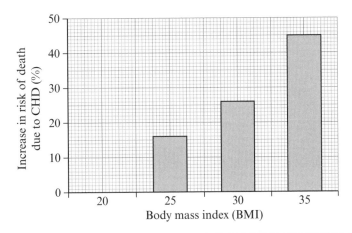

> Here, 'analyse' suggests you should do some calculations;
> the answer shows one sort of calculation you could do.

....

....

....

....   **(3 marks)**

3 Complete the table.

Waist measurement	Hip measurement	Waist : hip ratio
97 cm	112 cm	
61 cm	91 cm	
40 inches	46 inches	
52 cm		0.76
	138 cm	0.92

(5 marks)

Monosaccharides and disaccharides

1 Disaccharides can be split by

☐ **A** hydrolysis of glycosidic bonds

☐ **B** condensation of glycosidic bonds

☐ **C** hydrolysis of ester bonds

☐ **D** condensation of ester bonds.

(1 mark)

2 This diagram shows an α-glucose molecule. Draw a diagram to show the products formed when two α-glucose molecules join together to form maltose.

> There will be **two** products from this reaction.

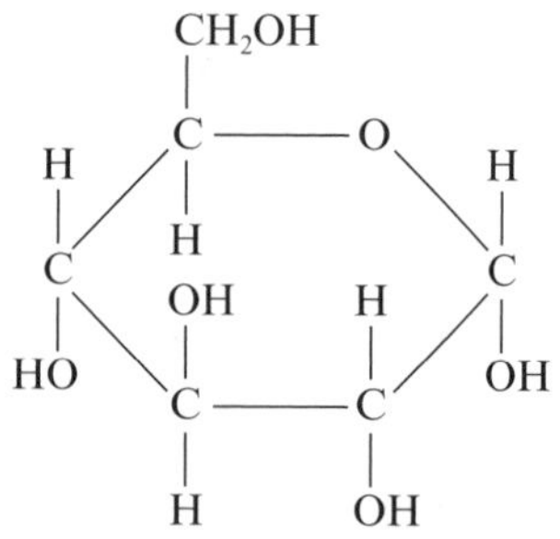

(3 marks)

3 Explain how the structures of glucose and sucrose relate to their roles in living things.

> This is an 'Explain' question, so it is not enough to state your point – you must also say why you have come to that conclusion.

....

....

....

....

....

....

(4 marks)

Carbohydrates – polysaccharides

>Guided〉 **1** Both glycogen and starch are energy-storage compounds in living organisms. Explain the features of these two molecules that adapt them to this role.

Both glycogen and starch are made from α-glucose molecules.

These glucose molecules are joined by ....

bonds ....

....

They are easily broken by ....

....

.... **(4 marks)**

2 The table shows some statements about polysaccharides. Indicate with a tick or cross in the box whether each statement is true or false for cellulose.

Statement	True	False
polymer of glucose		
molecule contains α- and β-glucose		
glycosidic bonds present		
molecules may have side branches		
molecules can form hydrogen bonds with adjacent molecules		

(5 marks)

3 Starch is a mixture of amylose and amylopectin. Explain why this allows it to provide both a quick and a slow release of glucose, and thus energy, after it is eaten.

....

....

.... **(2 marks)**

4 Compare and contrast the structure of amylopectin and glycogen.

....

....

....

.... **(3 marks)**

You must discuss both similarities and differences in 'compare and contrast' questions and look at both angles.

Lipids

1   This diagram illustrates part of the structural formula of a fatty acid with the formula $C_{10}H_{21}COOH$. Complete the diagram of the molecule.

> Note that there are **two** marks for this question and that the structural formula shows all the bonds.

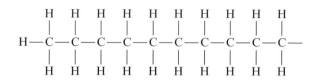

(2 marks)

2   This hydrocarbon chain is saturated. Redraw it to show a monounsaturated chain with the same number of carbon atoms.

> Note that this question also has **two** marks.

```
       H   H   H   H   H
       |   |   |   |   |
  H —  C — C — C — C — C —
       |   |   |   |   |
       H   H   H   H   H
```

(2 marks)

3   This table shows the concentration of eight fatty acids found in breast milk from two different groups of women. Analyse the data in order to compare the relative proportions of saturated and unsaturated fatty acids in the two groups of women.

Fatty acid	Number of double bonds in the hydrocarbon chain	Concentration of fatty acid/mg per g of breast milk	
		Vegans	Control group
Lauric	0	39	33
Myristic	0	68	80
Palmitic	0	166	276
Stearic	0	52	108
Palmitoleic	1	12	36
Oleic	1	313	353
Linoleic	2	317	69
Linolenic	3	15	8

....

....

....

....   **(3 marks)**

Good cholesterol, bad cholesterol

> Guided

1 The table shows the mean concentration of some types of lipid in the blood of people without a mutation in the gene coding for lipoprotein lipase and in the blood of people with the mutation. It is suggested that the mutation increases the risk of developing CVD. Analyse the data to explain why they do **not** support this suggestion.

Type of lipid	Mean concentration of lipid in blood/mg dm^{-3}	
	People without the mutation	**People with the mutation**
triglyceride	102	93
LDL cholesterol	121	111
HDL cholesterol	48	49
total cholesterol	186	179

The cholesterol levels in people with the mutation are not higher than

levels in people without the mutation. ....

....

....

.... **(3 marks)**

2 Explain how high blood cholesterol can influence the onset of CHD.

....

....

....

.... **(3 marks)**

3 Explain why LDLs should be maintained at a low level in the blood.

....

....

....

.... **(3 marks)**

Reducing the risk of CVD

1 Orange juice samples were heated for 25 minutes at five different temperatures. The vitamin C content was then determined using 0.1% DCPIP. The procedure was repeated five times at each temperature and a mean and standard deviation were calculated. The graph shows the results. Explain the effect of temperature on the vitamin C content of orange juice.

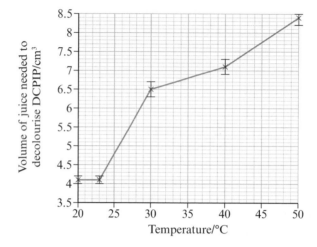

More vitamin C is needed to decolourise the DCPIP as temperature

increases ....

....

....

....

.... **(4 marks)**

2 There are a number of dietary changes which a person could make to lower their risk of CHD. Arrange the following list into those which should be increased and those which should be decreased in order to achieve this lowering of risk.

> A good way to do this would be in a two-column table.

saturated fat   fruits   oily fish   non-starchy polysaccharides

salt   cholesterol   unsaturated fats

....

....

.... **(2 marks)**

Medical treatments for CVD

Maths skills

1 Statins are drugs currently in use to help lower blood cholesterol levels. The table gives the results of some trials carried out to determine if statins reduce the risk of heart disease in men and women.

Drug used	Number of subjects involved		% of subjects who had heart attacks		% reduction in heart attacks
	Control group	Statin group	Control group	Statin group	
statin A	4502	4512	15.9	12.3	22.6
statin B	2223	2221	15.0	8.7	42.0
statin C	3301	3304	14.8	8.9	

(a) Calculate the % reduction in heart attacks after taking statin C.

> If you get one of these types of questions where answers are supplied, use them to check your answers. Also note that any percentage calculation is the same either way round. So, 25% of 34 is the same as 34% of 25. The first is easiest to do: 25% is $\frac{1}{4}$ so the answer is simply $\frac{34}{4}$.

Answer .... **(2 marks)**

(b) Suggest reasons for

(i) the number of subjects involved.

.... **(1 mark)**

(ii) the control group.

.... **(1 mark)**

2 (a) Explain why patients with CHD would take platelet inhibitory drugs, such as aspirin, and antihypertensives, such as beta blockers.

....

....

....

....

....

.... **(4 marks)**

(b) Give one risk of using **each** of these drugs.

....

.... **(2 marks)**

Daphnia heart rate

1 The graph shows the effect of caffeine concentration on the mean heart rate of *Daphnia*.

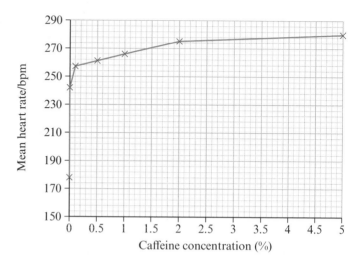

(a) It was concluded that there is a correlation between the two variables.
Explain what this means with reference to the graph.

....

.... **(1 mark)**

Maths skills

(b) A statistical test was carried out to assess the significance of the correlation between heart rate and caffeine concentration. The test used gave a correlation value of 1.00. The table shows the correlation values and significance levels for this test. Analyse the information in the table and the graph to explain the conclusion that can be drawn from this investigation.

Number of means	Significance level (p)				
	0.50	**0.20**	**0.10**	**0.05**	**0.01**
4	0.60	1.00	–	–	–
5	0.50	0.80	0.90	–	–
6	0.37	0.66	0.83	0.89	1.00
7	0.32	0.57	0.71	0.79	0.93
8	0.31	0.52	0.64	0.74	0.88
9	0.27	0.48	0.60	0.70	0.83
10	0.25	0.46	0.56	0.65	0.79

....

....

....

....

....

.... **(4 marks)**

(c) State and explain **one** ethical reason why the student chose to use *Daphnia* for this investigation.

.... **(2 marks)**

Exam skills

1   A student decided to investigate the effect of lactic acid on heart rate in water fleas (*Daphnia* sp.). *Daphnia* were placed in solutions of different concentrations of lactic acid, kept at 25 °C. Their heartbeats were observed and the heart rates were recorded. The procedure was repeated three times for each concentration.

(a) State **two** practical reasons why *Daphnia* was chosen for this investigation.

....

....   **(2 marks)**

(b) State **two** ethical considerations of using *Daphnia* in this investigation.

....

....   **(2 marks)**

(c) Plot a suitable graph to show the effect of lactic acid concentration on the mean heart rate of *Daphnia*. On your graph, include the standard deviations.

Lactic acid concentration/a.u.	Mean heart rate/bpm	Standard deviation
0	278	3.6
1	246	23.5
4	189	12.3
10	152	10.8
80	66	5.1

(5 marks)

In starred questions (*), structure your answer logically showing how the points you make are related to or follow on from each other. You need to select and apply relevant knowledge of biological facts or concepts to support the argument.

(d) *Analyse the data in the table and the graph you have plotted to explain the effect of lactic acid on *Daphnia* heart rate.

....

....

....

....

....

....   **(6 marks)**

Gas exchange

1 An investigation was carried out into the relationship between area to volume ratio and rate of diffusion. Agar blocks (cubes of equal sides) of different sizes were placed in a dye and the time taken for the dye to reach the centre of the block was recorded. The table shows the block size and the times taken.

Length of side/cm	Area of surface of cube/cm^2	Volume of cube/cm^3	SA/V	Time for whole block to become coloured/seconds
13	1014			380
10		1000		300
7			0.86	100
5	150			53
3				20

Here you need to know how to calculate area and volume of a cube from the length of a side. Simply cube it for volume and square it for area. Both of these can be done on a calculator either using the relevant button x^2 x^3 or simply multiplying the length by itself once (for area) or twice (for volume).

(a) Complete the table by adding the area, volume and SA/V.   **(5 marks)**

(b) Plot a graph of the relationship between SA/V and time taken.   **(4 marks)**

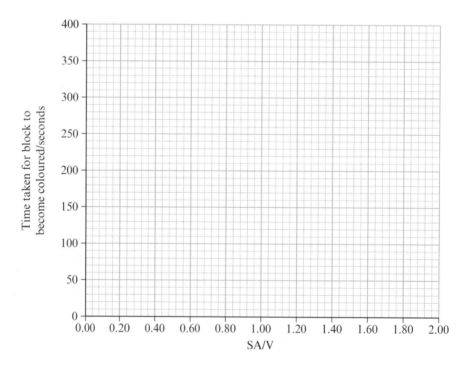

(c) Analyse the data in your graph to support the hypothesis that small organisms are able to carry out sufficient gas exchange to meet their needs without specially adapted gas exchange surfaces such as gills or lungs.

Describe the trend in the data first, then say how it affects the need for a gas exchange surface.

....

....

....

....   **(4 marks)**

The cell surface membrane

1 The centre of the cell membrane is made of a phospholipid bilayer. Explain why phospholipids form bilayers.

> This is all about which part of the phospholipid molecule is hydrophobic and which is hydrophilic.

....

....

....

.... **(3 marks)**

2 The photograph shows the membrane between two cells. Magnification is 1×10^6. Calculate the thickness of the membrane between A and B. Express your answer in standard form as a fraction of a metre and in nanometres. Use the magnification formula:

> Substitute what the image measures into the equation (the magnification of an object is what its image measures divided by what it actually measures) then convert to metres and nanometres. Remember a mm is 10^{-3} m and a nanometre is 10^{-9} m.

$$\text{magnification} = \frac{\text{size of image}}{\text{size of real object}}$$

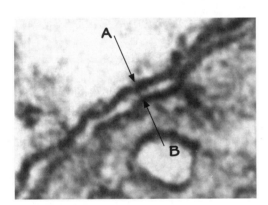

Answer: .... **(3 marks)**

Passive transport

1 An investigation was carried out into the permeability of a cell membrane in a number of different non-polar, organic molecules. The molecules differed in their size and their solubility in oil compared to their solubility in water. The higher the solubility, the more soluble the molecule is in oil compared to water. The graph shows the results of this investigation. The size of the circle drawn on the graph indicates the size of the molecule: the larger the circle, the larger the molecule.

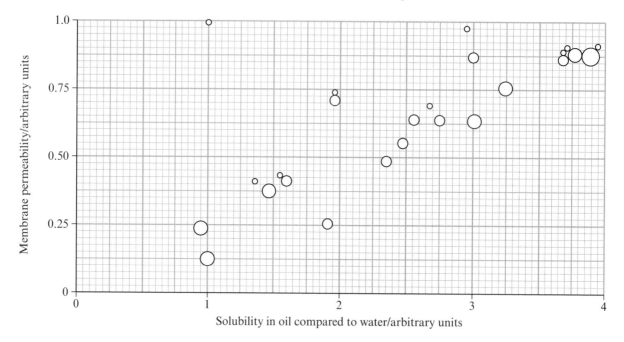

A student concluded that the data supported the fluid mosaic model of membrane structure. Analyse the data from this experiment to assess the validity of this conclusion.

> The data might support one aspect of the fluid mosaic model and not another. For example, the data might support the hydrophobic phospholipid bilayer but does it support the protein channels?

....

....

....

.... **(4 marks)**

2 The table shows some of the ways in which molecules can pass across membranes. Complete the table by writing correct or incorrect in each cell.

Process	Requires energy from respiration (ATP)	Requires a concentration gradient
passive diffusion		
facilitated diffusion		
osmosis		
active transport		

(4 marks)

Active transport, endocytosis and exocytosis

Guided   **1**   Compare and contrast transport across cell membranes by endocytosis and exocytosis.

Both processes involve the use of ....

Both processes also require ....

Endocytosis moves ....

but exocytosis transports ....   **(3 marks)**

Practical skills   **2**   An experiment is carried out to investigate movement across the membrane of a cell. The rate of movement of substances was followed at a variety of different concentrations of the transported substance. In one experiment, the membrane does not have specific channels for the substance. In another, there are carrier proteins; one is without a competitor and the other is with a competitor. The graph shows the results.

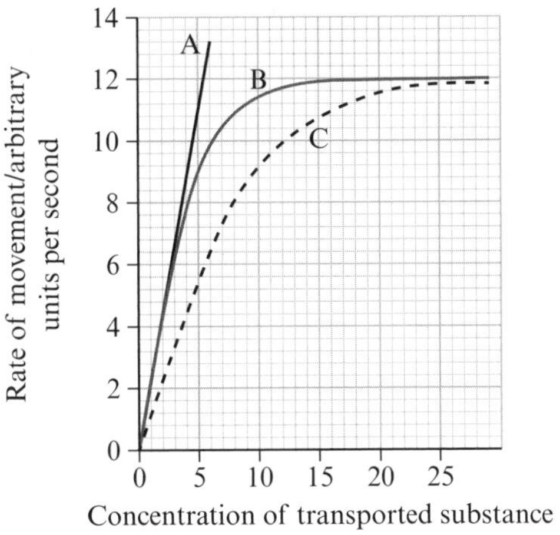

> A is linear so a straightforward relationship between concentration and rate of movement. Both lines B and C level off/reach a plateau, indicating that something is limiting the rate of movement. Since C levels off earlier, this suggests this is the line where the competitor is acting.

Which graph line shows the results for movement

(a) with no specific channel?   ....

(b) via a channel protein with no competitor for the movement?   ....

(c) via a channel protein with a competitor?   ....

Explain your answers.

....

....

....

....

....   **(5 marks)**

Practical on membrane structure

1 A piece of beetroot was placed in a tube containing 15 cm³ of water and left for 15 minutes. This was repeated for seven different concentrations of ethanol. A sample of the fluid around the beetroot was placed in a colorimeter to determine the intensity of red colouration of the fluid. The results are shown in the graph.

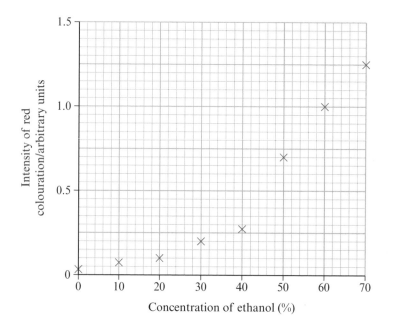

(a) State the colour of the filter that would have been used in the colorimeter.

.... **(1 mark)**

(b) State **two** biotic and **two** abiotic variables you need to keep constant in this investigation.

| Biotic will be something to do with the living material; abiotic with the non-living environment in which the investigation is carried out. |

....

....

....

.... **(4 marks)**

(c) Explain the result at 0% ethanol.

....

.... **(2 marks)**

(d) Explain the effect that increasing concentrations of alcohol have on beetroot membranes as shown in the graph.

....

....

....

.... **(3 marks)**

The structure of DNA and RNA

1   DNA consists of a double helix composed of two strands held together by

   ☐   **A**   hydrogen bonds between the bases of nucleotides.

   ☐   **B**   phosphodiester bonds between the bases of nucleotides.

   ☐   **C**   hydrogen bonds between the R side chains of amino acids.

   ☐   **D**   phosphodiester bonds between the R side chains of amino acids.   **(1 mark)**

> In looking at this question there are **two** ideas to think about: what DNA is made of and what types of bonds are used.

2   The following are simple drawings of deoxyribose, phosphate and a base.

   (a)   Using them, draw a nucleotide in the space below. Label your drawing.

> You must use the shapes given. The first step is to identify which shape relates to which unit, for example, deoxyribose is a pentose sugar. You then need to think about which is joined to which and where the bonds are.

(3 marks)

Guided   (b)   Clearly describe your drawing in part (a).

> You need to say where the bonds are and how they are formed.

The phosphate group is attached to the deoxyribose sugar

at carbon atom number five by a .... reaction and

....

....

....   **(3 marks)**

Protein synthesis – transcription

1 Name the substances X, Y and Z in the diagram.

X is ....

Y is ....

Z is ....

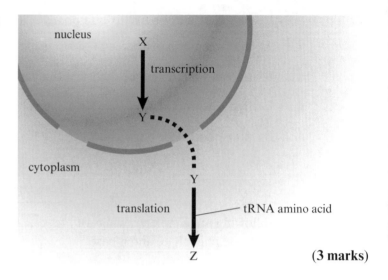

(3 marks)

2 Protein synthesis involves the process of transcription followed by translation. The diagram shows part of a messenger RNA (mRNA) molecule. It shows a sequence of seven bases.

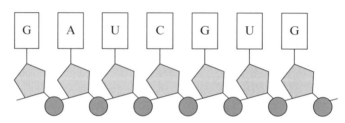

(a) Name the pyrimidine bases shown in this sequence.

....

.... **(1 mark)**

(b) Write down the sequence of bases on the strand of DNA that coded for this messenger RNA.

....

.... **(1 mark)**

(c) In some prokaryotic cells, the production of mRNA during transcription can occur at a rate of 50 bases per second. Calculate how long it would take to produce an mRNA molecule that codes for a protein containing 200 amino acids.

> Three bases code for one amino acid.

Time taken .... **(3 marks)**

Translation and the genetic code

1 Compare and contrast a gene and a codon.

....

....

....

....

> Make sure you link the characteristics of a gene with the same characteristics for the codon and give at least **one** similarity and **one** difference.

(3 marks)

2 Explain **one** advantage of degeneracy in the genetic code.

....

....

....

> In this case, 'explain' means that you need to describe what degeneracy is first. Also notice the bold 'one'; this means that only **one** advantage will gain any credit.

(2 marks)

⟩Guided⟩ **3** Identify the correct order of the four steps in DNA replication shown below.

A The enzymes DNA polymerase and DNA ligase join the nucleotides together.

B Hydrogen bonding links the two strands together.

C The two strands of DNA unwind and split apart.

D Free nucleotides line up along each strand, observing the complementary base pairing rules.

Step 1 is C   Step 2 is ......   Step 3 is ......   Step 4 is ......   **(2 marks)**

⟩Guided⟩ **4** Describe the processes of transcription and translation.

Transcription is when ....

....

....

Translation is when ....

....

....

....

....   **(5 marks)**

Amino acids and polypeptides

1 The bond between two amino acids is

☐ **A** a glycosidic bond.

☐ **B** a peptide bond.

☐ **C** a phosphodiester bond.

☐ **D** an ester bond.

(1 mark)

Guided

2 Draw the general structure of an amino acid.

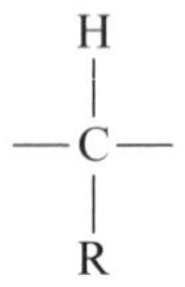

```
      H
      |
   — C —
      |
      R
```

> There is an amino group on one side of the central carbon and a carboxyl group on the other side.

(3 marks)

3 (a) The diagram shows parts of two amino acids. Complete it to show them joined together.

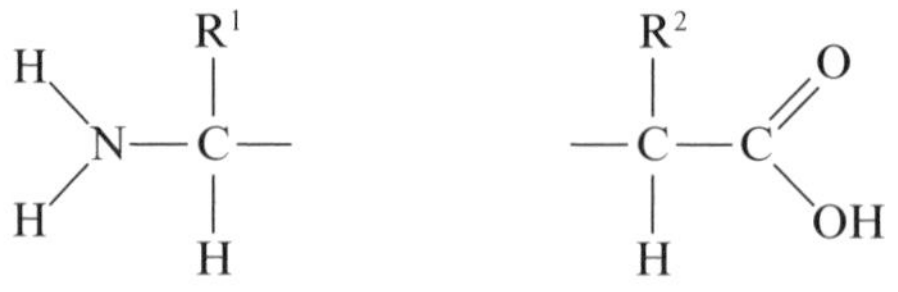

```
  H        R¹                    R²      O
   \       |                     |      ⫽
    N — C —               — C — C
   /       |                     |      \
  H        H                     H       OH
```

(2 marks)

(b) State the type of molecule you have drawn.

....

.... **(1 mark)**

(c) Are the two amino acids the same? Explain your answer.

> The answer is **no**.

....

.... **(1 mark)**

4 The diagram shows an amino acid. Which part makes all amino acids acidic?

☐ **A**

☐ **B**

☐ **C**

☐ **D**

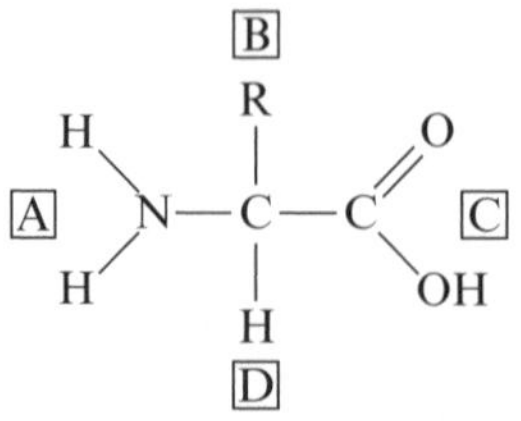

```
              B
              R
   H          |          O
    \         |         ⫽
  A  N — C — C
    /         |         C
   H          |          OH
              H
              D
```

(1 mark)

Folding of proteins

Guided 1 (a) The diagram shows the structure of an enzyme. Use the diagram to explain what is meant by primary structure, secondary structure and tertiary structure.

Note that the question asks for reference to the diagram so you must quote from it in your answer.

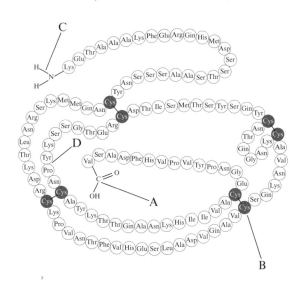

Primary structure is the sequence of ....

....

Secondary structure is the part of the structure stabilised by

....

....

Tertiary structure is ....

....

.... **(6 marks)**

(b) State the names of the groups labelled A and C.

A is ....   C is ....   **(2 marks)**

2 State the name and type of bond found between the two cysteines at B.

....

.... **(2 marks)**

3 Describe what might happen to a polypeptide to make it into a functional protein such as an enzyme.

You have to distinguish between a polypeptide and a protein and make **two** points for **two** marks.

....

.... **(2 marks)**

Haemoglobin and collagen

>Guided>

1 Compare and contrast fibrous proteins with globular proteins.

> For 'contrast', make sure your answer makes it clear what the differences are by writing your answer as a comparison. An example is given. For 'compare', you need to point out something about them that is similar.

Fibrous proteins are long chains whereas globular proteins are

spherical. ....

....

....

....   **(3 marks)**

2 Describe the structure of collagen.

> For 'describe' questions, you do **not** need to make a judgement or explain how or why, you only need to describe what is requested.

....

....

....

....

....

....

....   **(5 marks)**

3 Explain how the structure of haemoglobin is adapted to its function of picking up oxygen in the lungs and releasing it at tissues.

....

....

....

....

....

....

....

....   **(5 marks)**

Enzymes

1   *Amoeba* is a small single-celled animal which engulfs bacteria for food. The bacteria are digested inside the cell by enzymes. Which of the following types are these enzymes examples of?

   ☐   **A**   extracellular

   ☐   **B**   intercellular

   ☐   **C**   intracellular

   ☐   **D**   extercellular   **(1 mark)**

2   (a)   Caffeine is converted in the liver by an enzyme complex called cytochrome P450 oxidase. Three products are formed as shown in the diagram. Explain how this shows that the cytochrome P450 oxidase complex must be more than one enzyme.

> Each of the products has the same general formula, $C_7H_8N_4O_2$, so you need to think about how they differ.

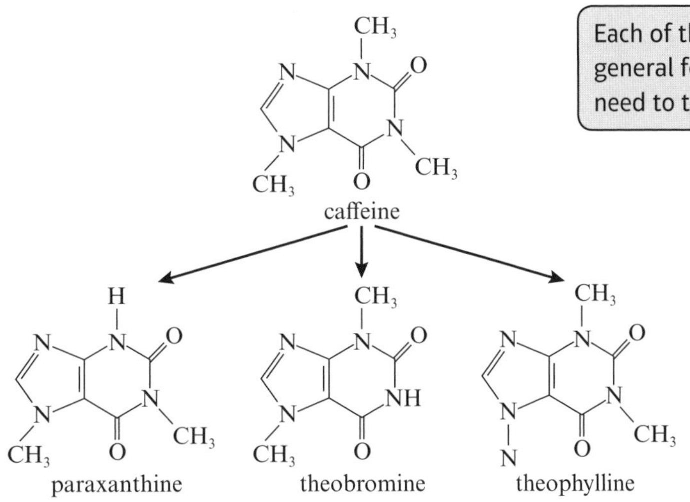

caffeine

paraxanthine   theobromine   theophylline

....

....

....

.... **(3 marks)**

   (b)   Paraxanthine : theobromine : theophylline are produced in the ratio 84 : 12 : 4. Assuming 50 milligrams of paraxanthine was produced after a cup of coffee, calculate the mass of theobromine **and** theophylline this person would produce. Show your working.

> Notice that 12 divides into 84 seven times.

theobromine .... mg

theophylline .... mg   **(4 marks)**

Activation energy and catalysts

1 Which of the following statements is true of enzymes?

   ☐ **A**   They cause reactions to happen by raising the activation energy.

   ☐ **B**   They increase the rate of reaction by lowering the activation energy.

   ☐ **C**   They increase the rate of reaction by raising the activation energy.

   ☐ **D**   They cause reactions to happen by lowering the activation energy.   **(1 mark)**

> There are **two** parts to each answer. For each part, decide if it is correct then see which statement has both parts correct.

2 Enzymes lower the activation energy of a reaction by providing an alternative pathway for reactions. Explain how they do this.

> To answer this fully you need to think about enzymes both breaking **and** making bonds and what happens in the active site when the substrate or substrates are in there.

....

....

....

....   **(3 marks)**

3 Look at this energy diagram of a reaction with and without an enzyme. Explain what A, B and C show.

A ....

....

B ....

....

C ....

....   **(3 marks)**

Reaction rates

1 A student is asked to design an experiment to look at the effect of pH on the initial rate of the enzyme reaction in which hydrogen peroxide is broken down into water and oxygen. She proposes a procedure in which she times how long it takes for $100 \, cm^3$ of oxygen to be produced at each of five pH values. The rate is then calculated by finding the reciprocal of the time taken (1/time). Explain what is wrong with her procedure for calculating the rate.

> Think about what happens to the amount of substrate as the reaction proceeds.

....

....

....

....

....

....   **(4 marks)**

2 The graphs show the results of two experiments to determine the initial rate of two different enzyme-catalysed reactions. In experiment 1, a protease enzyme breaks down a suspension of a protein, which was initially cloudy. In experiment 2, a solution of glucose phosphate is converted into starch using starch synthetase enzyme. In both cases, the course of the reaction over time is followed using a colorimeter. Explain why the graphs are different shapes.

> Think about the products of each experiment and what effect they will have on the cloudiness of the solution as the reaction progresses.

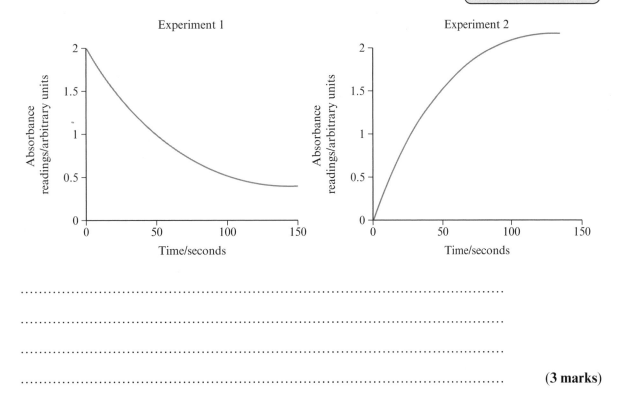

....

....

....

....   **(3 marks)**

Initial rates of reaction

1 Read this method for a practical designed to find the optimum temperature of an enzyme:

> Pipette 2 cm³ of 1% protein solution into a cuvette. Pipette 2 cm³ of 1% protease solution into the same cuvette. Mix thoroughly and immediately put into a colorimeter and start the stop clock. Measure and record the absorbance at suitable time intervals until there is little change. Repeat the procedure using a range of different temperatures ensuring that other conditions are unchanged.

(a) State the dependent variable in this experiment.

....   **(1 mark)**

(b) The method, says 'ensuring that other conditions are unchanged'. Explain what this means and how it would be achieved with reference to this experiment.

> It is always important to read the stem of a question very carefully. Here, you need to do that to answer the question.

....

....

....

....   **(3 marks)**

(c) Explain how you would use the data to find the initial rate of reaction for each temperature.

....

....

....

....   **(3 marks)**

2 State how temperature would be controlled in an experiment to look at the effect of pH on the same enzyme.

> You must be very precise in answers to A level questions, so 'in a water bath' is not sufficient detail.

....

....

....   **(2 marks)**

Exam skills

1 (a) The graph shows the effect of the enzyme lipase on the initial rate of
breakdown of lipid at different concentrations. Analyse the data to explain
the effect of lipase on this reaction.

> The word 'Analyse'
> means that you need
> to comment on the
> data and then relate
> your comments to
> the situation being
> discussed to make a
> judgement.

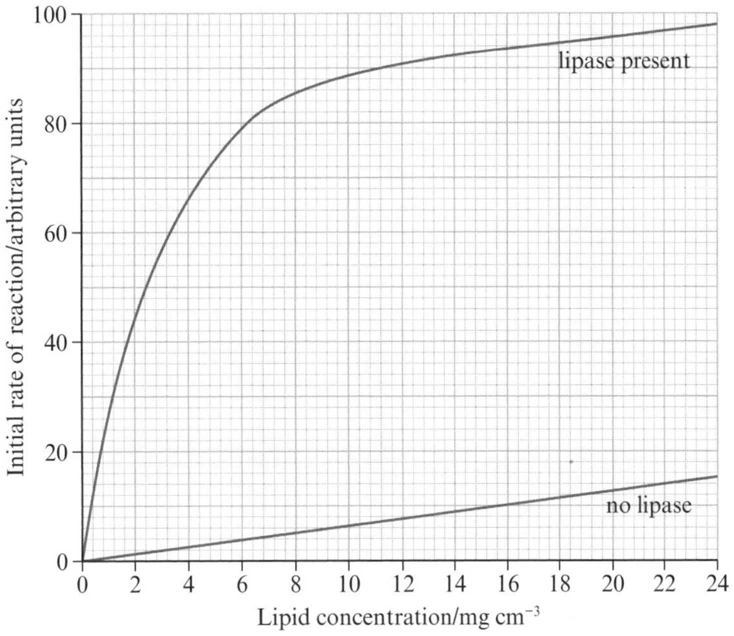

....

....

....

....   **(3 marks)**

(b) (i) Name the bond broken by the lipase enzyme.

....   **(1 mark)**

(ii) Name **two** products formed from the breakdown of triglycerides by the
lipase.

....

....

....   **(2 marks)**

(iii) Explain how the breakdown of triglycerides would affect the pH of the
reaction mixture.

....

....

....

....   **(3 marks)**

Had a go ☐ Nearly there ☐ Nailed it! ☐

DNA replication

1 What is the correct order of the four steps in DNA replication?

P The enzymes DNA polymerase and DNA ligase join the nucleotides together.

Q Hydrogen bonding links the two strands together.

R The two strands of DNA unwind and split apart.

S Free nucleotides line up along each strand, observing the complementary base pairing rules.

Step 1 is .... Step 2 is ....
Step 3 is .... Step 4 is .... **(2 marks)**

2 If an organism is supplied with nucleotides containing a heavy isotope and allowed to replicate once using these, what will be the mass of the new DNA formed?

☐ **A** normal

☐ **B** heavy

☐ **C** between normal and heavy

☐ **D** no DNA could be made with this isotope **(1 mark)**

3 The diagram shows part of the process of DNA replication.

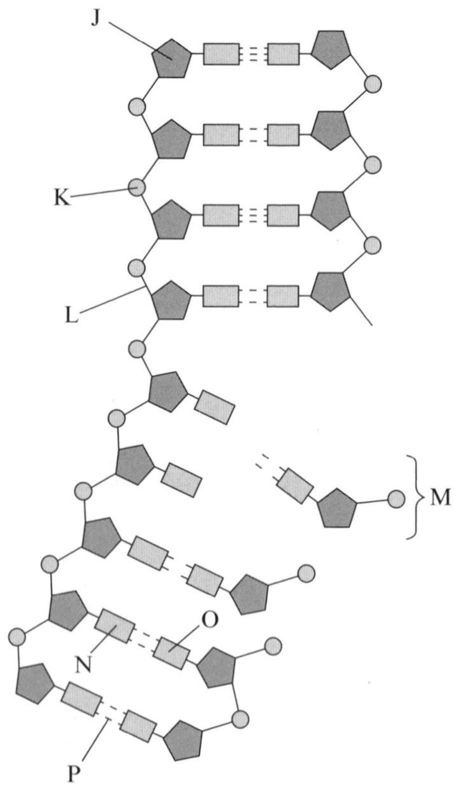

(a) Name the structure labelled J.

.... **(1 mark)**

(b) Name the structure labelled K.

.... **(1 mark)**

(c) Name the bond labelled L.

.... **(1 mark)**

(d) Name the structure labelled M.

.... **(1 mark)**

(e) Name O on the new DNA molecule, assuming the base labelled N on the parent DNA molecule is adenine.

.... **(1 mark)**

(f) Name the bond labelled P.

.... **(1 mark)**

Evidence for DNA replication

1 Meselson and Stahl carried out an experiment that demonstrated the semi-conservative replication of DNA.

(a) Explain the meaning of the term semi-conservative replication.

....

....

.... **(2 marks)**

(b) Meselson and Stahl grew bacteria in a medium containing one isotope of nitrogen. These bacteria were then transferred to a second medium containing a different isotope of nitrogen. The DNA extracted from the bacteria was then separated according to its density. Draw bands on tubes B and C to show the results of this experiment.

DNA from bacteria grown in first medium

DNA from bacteria allowed to replicate once in second medium

DNA from bacteria allowed to replicate twice in second medium

A   B   C

(2 marks)

(c) *A source said that Meselson and Stahl's experiments 'supported the accepted theory of replication of DNA and refuted competing theories'. Explain what this means.

> In starred questions (*), structure your answer logically showing how the points you make are related to or follow on from each other. You need to select and apply relevant knowledge of biological facts or concepts to support the argument.

....

....

....

....

....

....

.... **(6 marks)**

Mutation

Guided 1 Krabbe disease is caused by mutations in a gene called GALC. This leads to a lack of the enzyme galactocerebrosidase. Explain how a mutation in the GALC gene could result in a change in the protein (which would normally become the active enzyme galactocerebrosidase) and how this might stop the formation of an active enzyme.

A gene is a sequence of bases that codes for a sequence of amino

acids in a protein. A mutation is a change in that sequence, so ....

....

....

....

....

....   **(4 marks)**

2 A blood disorder called sickle cell anaemia is caused by a mutation of the gene that codes for haemoglobin. The diagram shows the DNA, mRNA and resultant amino acids for the relevant section of the normal gene and the mutated one.

> Look at the amino acid sequence and identify what has been changed then backtrack to the base sequence to see what has happened.

DNA: C-A-C-C-T-G-G-A-C-T-G-A-G-G-A-C-T-C-C-T-C   DNA: C-A-C-C-T-G-G-A-C-T-G-A-G-G-A-C-A-C-C-T-C
RNA: G-U-C-G-A-C-C-U-G-A-C-U-C-C-U-G-A-G-G-A-G   RNA: G-U-C-G-A-C-C-U-G-A-C-U-C-C-U-G-U-G-G-A-G

<u>Val</u> <u>His</u> <u>Leu</u> <u>Thr</u> <u>Pro</u> <u>Glu</u> <u>Glu</u>   <u>Val</u> <u>His</u> <u>Leu</u> <u>Thr</u> <u>Pro</u> <u>Val</u> <u>Glu</u>

normal   mutated

Which of the following describes the gene mutation for sickle cell anaemia?

☐ **A** insertion mutation   ☐ **C** translocation mutation

☐ **B** deletion mutation   ☐ **D** substitution mutation   **(1 mark)**

3 The CFTR protein is either not synthesised or has reduced or no functionality in a condition called cystic fibrosis. How might a mutation lead to

(a) no synthesis of CFTR?

....

....   **(2 marks)**

(b) synthesis of CFTR but with reduced functionality?

....

....   **(2 marks)**

Classical genetics

1   In the four-o'clock plant (*Mirabilis jalapa*) a cross between
pure breeding red-flowered individuals and pure breeding
white-flowered ones gives all pink offspring. Crosses of
these give white-, pink- and red-flowered varieties. Calculate
the ratio of white to pink to red in such a cross. Show your
reasoning with a cross diagram.

> You will need to choose symbols
> for the two alleles. Since neither
> is dominant, the letters should
> be different but the same case.

Ratio = ...   **(4 marks)**

2   This pedigree diagram is for a condition called galactosaemia.

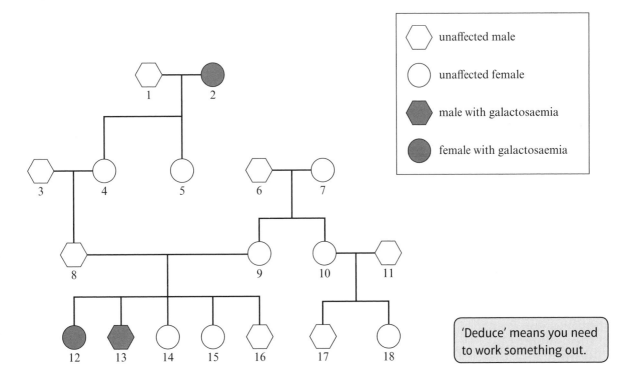

unaffected male

unaffected female

male with galactosaemia

female with galactosaemia

> 'Deduce' means you need
> to work something out.

Deduce the probability that the next child of 8 and 9 will have galactosaemia.

....

....

....

....

....

....   **(4 marks)**

Cystic fibrosis symptoms

Guided 1 People with cystic fibrosis produce sticky mucus. Explain how sticky mucus can reduce the rate of gas exchange in the lungs.

Gas exchange depends on diffusion and rate of diffusion depends on

....

....

.... **(2 marks)**

2 Explain why women with cystic fibrosis may find it difficult to become pregnant.

....

....

.... **(2 marks)**

3 Three treatments for the symptoms of cystic fibrosis are physiotherapy, digestive enzyme supplements and antibiotics. Explain how each relieves the symptoms of this disorder.

> Make sure you link each treatment with its explanation.

Physiotherapy ....

....

Digestive enzyme supplements ....

....

Antibiotics ....

.... **(3 marks)**

4 Explain why diabetes can be associated with cystic fibrosis.

> You need to remember what diabetes is.

....

....

.... **(2 marks)**

Genetic screening

1 Discuss the issues that carrier couples might consider when
deciding whether or not to have chorionic villus sampling (CVS).

In such questions, try to give
both sides: in this case, the
problems and the benefits.

There is a 1–2% risk of miscarriage so the couple may lose the baby.

....

....

....   **(2 marks)**

2 There are two prenatal tests for cystic fibrosis: amniocentesis with a 0.5–1.0%
risk of miscarriage, and chorionic villus sampling where the risk of miscarriage
is greater (1–2%). Explain why, despite the greater risk, chorionic villus sampling
may be regarded as preferable to amniocentesis.

CVS can be carried out earlier than amniocentesis ....

....

....   **(2 marks)**

3 Explain how pre-implantation genetic diagnosis is performed.

....

....

....   **(2 marks)**

4 Discuss **one** ethical issue and **one** social issue relating to the use
of pre-implantation genetic diagnosis.

Do not just discuss any two
issues. Try to make sure **one**
is ethical and **one** is social.

....

....

....

....

....

....   **(4 marks)**

Exam skills

1   The diagram shows the base sequence on a short section of DNA consisting of 12 mononucleotides. This base sequence contains the genetic code for a short section in the primary structure of a polypeptide.

section of DNA	A	A	T	A	A	C	C	A	G	T	T	T

amino acids   leucine   leucine   valine   lysine

(a)   Name each of the bases represented by the letters, A, C, G and T in the diagram.

A is .... C is .... G is .... T is .... **(1 mark)**

(b)   Using the sequence shown in the diagram, explain the meaning of each of these terms: triplet, non-overlapping and degenerate code.

....

....

....

.... **(3 marks)**

(c)   Which of the following names two of the components, other than the bases, that form part of each mononucleotide in this sequence?

☐ **A**   deoxyribose and nitrate

☐ **B**   deoxyribose and phosphate

☐ **C**   ribose and nitrate

☐ **D**   ribose and phosphate **(1 mark)**

(d)   Transcription of this section of DNA forms a complementary strand of mRNA.

Describe how translation of this mRNA synthesises part of a polypeptide molecule.

> In questions of this kind, where a context is provided, it is important to write your answer in terms of that context. In the case of (d), 'this mRNA' is the key phrase. Look for context in (b) as well.

....

....

....

....

....

.... **(5 marks)**

Exam skills

1 Enzymes are proteins that speed up chemical reactions within living organisms. The graph shows the effect of changing enzyme concentration on the rate of reaction.

In this case, you are given the observation and simply asked to explain it. Sometimes, 'explain' questions will require you to describe what is going on and then explain it.

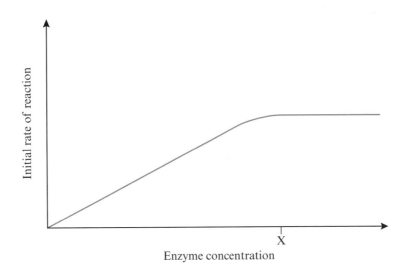

Enzyme concentration

(a) Explain why increasing the enzyme concentration above point X on the graph does **not** increase the rate of the reaction further.

....

....

.... **(2 marks)**

Practical skills

(b) Describe the practical procedures that could have been used to obtain the results shown on the graph.

This is a core practical and you should know the details thoroughly. In this case, as the graph is generic you can write your answer in terms of a practical which you have carried out.

....

....

....

....

....

....

.... **(5 marks)**

Prokaryotes

1 The diagram shows a prokaryotic cell. Complete the labelling.

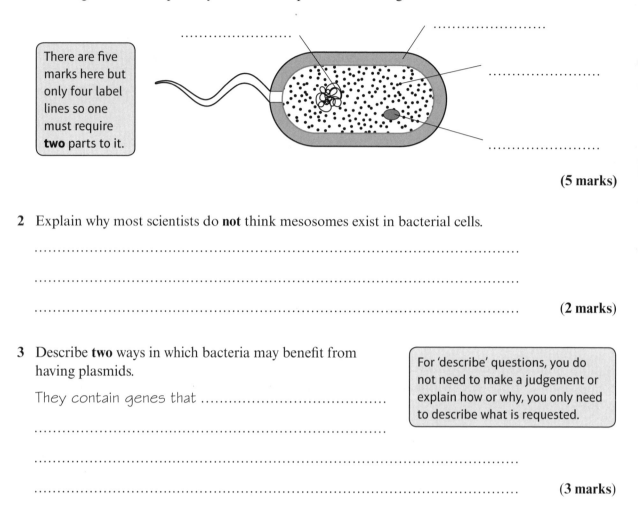

There are five marks here but only four label lines so one must require **two** parts to it.

(5 marks)

2 Explain why most scientists do **not** think mesosomes exist in bacterial cells.

....

....

....   **(2 marks)**

Guided   3 Describe **two** ways in which bacteria may benefit from having plasmids.

They contain genes that ....

....

....

....   **(3 marks)**

For 'describe' questions, you do not need to make a judgement or explain how or why, you only need to describe what is requested.

4 The diagram shows a mitochondrion. Two of the features labelled are typical of prokaryotes. Place a tick in each of the **two** boxes that correctly identify these features.

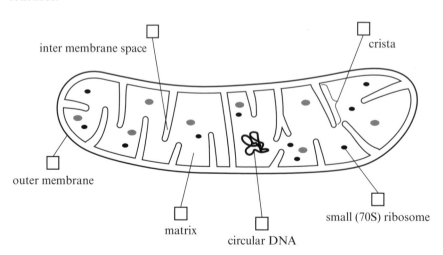

inter membrane space   crista

outer membrane

matrix   circular DNA   small (70S) ribosome

(2 marks)

Had a go ☐   Nearly there ☐   Nailed it! ☐

Eukaryotes

 1 A student wrote a description of each of the organelles seen in an electron micrograph of an animal cell. Complete the student's table with the names of the organelles and a brief statement of what they do.

> The function is what the organelle does; two have been done for you.

Description	Name	Function
a large organelle with a double envelope with pores through it		stores DNA
a branching series of channels studded with small, roughly spherical structures		
quite large oval organelles with folded membranes inside		
a pair of cylindrical structures at right angles to each other		make the spindle fibre in cell division

(4 marks)

2 Complete the table for the three cell types using the list below the table.

> Do not feel that **all** boxes have to have an entry.

Animal cell only	Plant cell only	Animal and plant	Bacteria only	All three cell types

(3 marks)

> mitochondria   ribosomes   smooth endoplasmic reticulum (sER)
>
> DNA in a nucleus   centrioles   cell surface membrane

 3 Plant cells contain some organelles **not** found in animals. Describe what these are and their functions.

The organelles not found in animals are .... where photosynthesis

occurs, a vacuole where .... and .... are stored. Outside

the plant cell surface membrane is a ....

(4 marks)

Electron micrographs

1   The photograph shows part of an animal cell as seen through an electron microscope.

(a) Name structures A to D.

> Do **not** use abbreviations in your answers to this type of question.

A   ....

B   ....

C   ....

D   ....

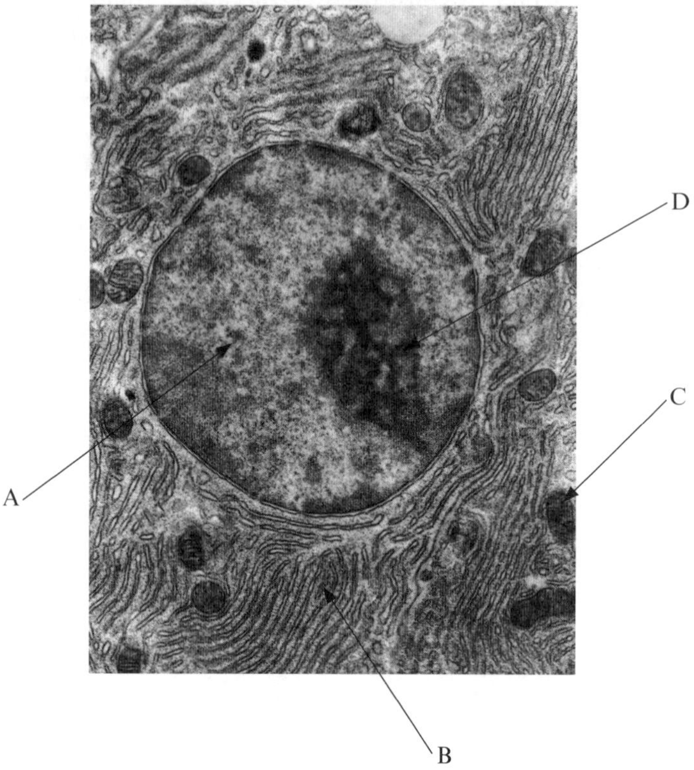

(2 marks)

(b) Structure B has small granular structures on its surface, too small to be seen at this magnification. State the name of these structures.

....   **(1 mark)**

(c) State the name of the structure that looks like structure B but has no granules.

....   **(1 mark)**

(d) Draw a lysosome as it would appear under the electron microscope.

(1 mark)

Protein folding, modification and packaging

Guided

1 During an infection some white blood cells make glycoproteins which become part of their cell surface membranes. To make glycoproteins, the white blood cells must first synthesise proteins on the surface of their rough endoplasmic reticulum (rER). Explain how these newly-made proteins end up as glycoproteins on the cell surface membrane.

The protein is released from ribosomes and enters the lumen of the rER.

In here, ...

....

....

....

.... **(4 marks)**

2 This cell was supplied with a pulse of radioactively-labelled amino acids followed by non-radioactive ones. Label it with numbers **1–5** to show the order of where you would expect to see radioactivity over the following few hours, where **1** is the first place radioactivity would be found.

Make sure you include clear and accurate labels and use a ruler if necessary.

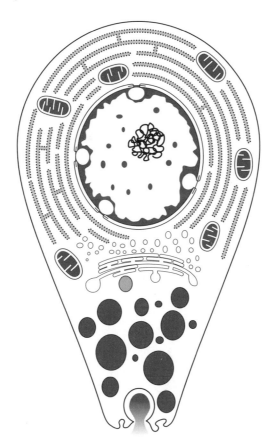

(5 marks)

Sperm and eggs

1   The diagram shows a human sperm and human egg.

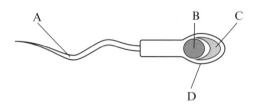

human sperm cell

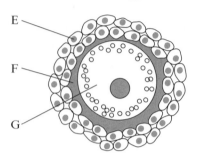

human egg

Write the correct letter or letters next to the statements below.

A site containing an enzyme that digests the zona pellucida   ....

A site with a haploid number of chromosomes   ....

A site containing mitochondria   ....

The zona pellucida   ....   **(4 marks)**

> Note that there are more letters than answers so make sure that you know exactly where the labels are going to.

⟩Guided⟩   2   Complete the table about mammalian sperm and eggs using an X to indicate whether sperm, eggs, both sperm and egg, or neither contain the feature.

> As well as their specialised features, sperm and eggs are eukaryotic cells like any other.

Feature	Egg only	Sperm only	Sperm and egg	Neither sperm nor egg
mitochondria			X	
DNA				
cortical granules				
membrane				
cell wall				
diploid nucleus				
mid-piece				

(7 marks)

Genes and chromosomes

1 Which of the following is true if an allele is found only on the X chromosome?

The allele can never pass from

☐ **A**   a woman to her daughter.

☐ **B**   a woman to her granddaughter.

☐ **C**   a man to his son.

☐ **D**   a man to his grandson.   **(1 mark)**

2 In fruit flies, vestigial wings (v) is recessive to normal wings (V) and black body (b) is recessive to normal body (B). These two genes, one for wing length and one for body colour, are linked.

(a) Draw a genetic diagram to show the offspring you would expect and the phenotypic ratio if a fly of genotype VvBb was crossed with another of the same genotype. The parents of both of these flies were VVBB and vvbb.

> A Punnett square is a good way to do this. Start with VvBb × VvBb.

(4 marks)

(b) Explain how the ratio you deduced in part (a) would change if linkage were **not** complete, that is, if there was crossing over.

....

....

....

.... **(3 marks)**

Meiosis

> **Guided**

1   Diagrams A and B show cells from the same organism. Both cells are in the same stage of nuclear division. One cell is undergoing mitosis and the other cell is undergoing meiosis. Explain which cell is undergoing meiosis.

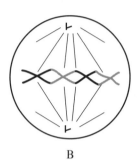

A   B

It is cell A because ....

....

....

....   **(3 marks)**

2   One way in which meiosis increases genetic variation is through crossing over.

The diagram shows a pair of homologous chromosomes during meiosis. They are positioned next to each other but crossing over has not yet occurred. Complete the diagram on the right to show these chromosomes after crossing over has occurred.

> Both will have some grey and some white.

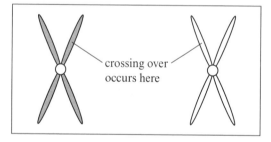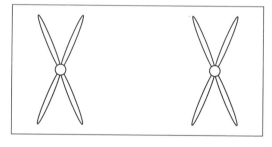

(1 mark)

3   Individuals who are heterozygous for two different linked genes are crossed. Mendel would have predicted one result from this cross. The theory that genes are on the same chromosome would give a different result. Explain why the actual result would be different from either of these predictions.

> Work out what the Mendelian result would be and what the linked result would be first.

....

....

....

....

....   **(4 marks)**

The cell cycle

1 Describe the events that occur in mitosis from the start of prophase
   up to the end of anaphase.

   ....

   ....

   ....

   ....

   ....

   ....

   ....

> This is an extended writing question so think carefully about the detail, plan it by listing the stages (prophase, metaphase, anaphase and telophase), then describe each one in order.

(4 marks)

2 The photograph shows cells in various stages of the cell cycle.

   (a) Name the stages shown by A and B.

   A is ....

   B is ....

> The chromosomes are on the equator in A but at the poles in B; no new nuclear membranes are being made as yet in either.

(2 marks)

   (b) Explain which part of the cell cycle is shown by C.

   ....

   ....

   ....

> This is an 'explain question' so it is not enough to say which part of the cell cycle; you must say why you have come to that conclusion.

(2 marks)

Mitosis

1   The flowchart shows the stages in the preparation of a root tip squash to observe cells in the cell cycle.

Procedure	**Reason**
Cut final 5 mm of a root tip from a plant such as an onion.	
Add stain.	
Add acid.	
Gently break open on a microscope slide and squash carefully.	
Warm, then look at slide on a microscope.	

(a) Complete the flowchart with reasons for the procedure.   **(5 marks)**

> At stage 5 you will need **two** reasons as there are two procedures.

(b) Name a suitable stain at the second stage.

.... **(1 mark)**

2   After a zygote is formed during fertilisation, cell division occurs.

(a) Calculate the number of cells that will be present in the embryo after the first four divisions of the zygote.

> Think about how many new cells are formed in every division.

.... **(1 mark)**

(b) What kind of cell division is involved in the growth of the embryo?

.... **(1 mark)**

> **Guided**

3   Greenfly reproduce asexually throughout the summer but sexually in the autumn ready for the winter. Give reasons why.

Conditions in summer are stable and suitable for greenfly so no

need to generate .... In winter and the following spring,

conditions might change so ....

.... **(2 marks)**

Exam skills

1   Yeast reproduces asexually by a process called budding. When the bud is big enough it separates from the original yeast cell. Explain the role of the cell cycle in the process of budding in yeast.

> Questions like this must be answered in the context given so you need to write about every stage of the cell cycle and how it is involved in this particular process.

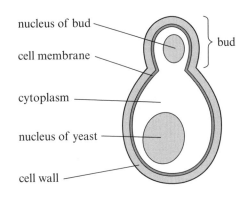

nucleus of bud

cell membrane

cytoplasm

nucleus of yeast

cell wall

bud

....

....

....

....   **(3 marks)**

2   Compare and contrast the first and second divisions of meiosis.

> You must discuss both similarities and differences in 'compare and contrast' questions and look at both angles.

....

....

....

....

....

....   **(4 marks)**

3   Explain how mutation differs from crossing over in the way that genetic variation is generated.

....

....

....

....   **(3 marks)**

Stem cells and cell specialisation

> **Guided**

1 Stem cells can differentiate into specialised cells and tissues. There are about 23 000 genes in a human body cell. The table below shows the number of genes that have not been switched off in three different cells, A, B and C. Analyse this information to support the conclusion that C is a totipotent stem cell whereas A and B are not.

Cell	Number of genes that have not been switched off
A	11 000
B	18 000
C	23 000

In cell C, all genes are potentially active as no genes are switched ....

....

....

....

....   **(3 marks)**

2 Scientists have successfully used stem cells to reverse paralysis due to spinal injury in dogs.

(a) Explain why stem cells can be used to reverse paralysis.

....

....

....   **(2 marks)**

> **Practical skills**

(b) A trial experiment was carried out on a dog that was paralysed due to a spinal injury. Stem cells were obtained from the dog and cultured to increase their numbers. They were injected into the injury site of the dog. Explain why stem cells were taken from this dog and not from another dog.

....

....

....   **(2 marks)**

(c) A further investigation was carried out on 34 dogs with spinal injuries. Some had stem cells injected into the site of the spinal injury. The others were injected with neutral fluid containing no stem cells. Explain why some of the dogs were given a neutral fluid instead of stem cells.

....

....

....   **(2 marks)**

Gene expression

1   In the lac operon system, lactose is the inducer molecule. Which of the following is true of the inducer in this system?

☐   **A**   It combines with the operator region and activates operons.

☐   **B**   It combines with repressor proteins and inactivates them.

☐   **C**   It combines the beta galactosidase gene and activates it.

☐   **D**   It directly activates RNA polymerase.   **(1 mark)**

2   Explain how acetylation and methylation are involved in gene expression.

> Make sure you can distinguish between these two similar words and their effects, one on DNA and one on histones.

....

....

....

....

....

....   **(4 marks)**

3   Explain what is meant by the terms tissue and organ.

....

....

....

....

....

....   **(4 marks)**

4   An operon consists of promoter, operator, regulator and structural genes. Describe what each of these genes does.

> For 'describe questions', you do not need to make a judgement or explain how or why, you only need to describe what is requested.

....

....

....

....

....

....   **(4 marks)**

Nature and nurture

1   The phenotype of an organism is affected by its genotype and its environment.

The table shows the mean difference in height and mass from a study on human identical and non-identical twins. Each pair of twins was brought up together. Analyse the data in the table to describe the effects that genotype and the environment have on the phenotype.

Phenotype	Types of twins	
	Identical	Non-identical
mean height difference/cm	1.7	4.4
mean mass difference/kg	1.9	4.6

....

....

....

....   **(3 marks)**

2   Rats that were very good at running through mazes (maze-bright) were allowed to breed to produce many more maze-bright rats. When they were young, they were split into three groups and each group was raised in one of three conditions as shown in the graph. The same was done with maze-dull rats. A rat was placed at the start of a maze and the number of errors it made was recorded as it ran through the maze. This was repeated using many rats, both maze-bright and maze-dull. The graph shows the results. Analyse the data to explain the effect of genes and environment on rat maze-running behaviour.

> The word 'analyse' means that you need to comment on the data and then relate your comments to the situation being discussed to make a judgement.

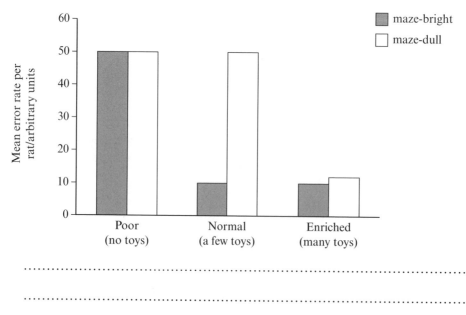

....

....

....

....

....

....   **(4 marks)**

Continuous variation

1 Flowers of some plants show continuous variation in colour due to polygenic inheritance rather than environmental factors. Explain what is meant by the term polygenic inheritance.

....

.... **(2 marks)**

2 An organism has five genes which affect height. A, B, C, D and L.
A, B, C and D each add 3 cm to height, 6 cm when two are present. The recessive alleles of these genes, a, b, c and d, each add nothing. Gene L has no recessive allele and always adds 40 cm.

Maths skills

(a) Calculate how tall an individual with genotype AABBCCDDLL and an individual with genotype aabbccddLL would be. Show your working in each case.

> In maths problems it is always important to write out each step to keep track of what you are doing and to allow others to follow your logic.

Answer: .... **(2 marks)**

(b) The following cross was made:

AABbCCDDLL × aaBBccDDLL

AbCDL   aBcDL

AaBbCcDDLL   HEIGHT IS 55 cm

AaBbCcDDLL × AaBbCcDDLL

	ABCDL	ABcDL	AbCDL	AbcDL	aBCDL	abCDL	aBcDL	abcDL
ABCDL	AABBCCDDLL 64	AABBCcDDLL 61	AABbCCDDLL 61	AABbCcDDLL 58	AaBBCCDDLL 61	AaBbCCDDLL 58	AaBBCcDDLL 58	AaBbCcDDLL 55
ABcDL	AABBCcDDLL 61	AABBccDDLL 58	AABbCcDDLL 58	AABbccDDLL 55	AaBBCcDDLL 58	AaBbCcDDLL 55	AaBBccDDLL 55	AaBbccDDLL 52
AbCDL	AABbCCDDLL 61	AABbCcDDLL 58	AAbbCCDDLL 58	AAbbCcDDLL 58	AaBbCCDDLL 58	AabbCCDDLL 55	AaBbCcDDLL 55	AabbCcDDLL 52
AbcDL	AABbCcDDLL 58	AABbccDDLL 55	AAbbCcDDLL 55	AAbbccDDLL 52	AaBbCcDDLL 55	AabbCcDDLL 52	AaBbccDDLL 52	AabbccDDLL 49
aBCDL	AaBbCCDDLL 58	AaBBCcDDLL 58	AaBbCCDDLL 58	AaBbCcDDLL 55	aaBBCCDDLL 58	aaBbCCDDLL 55	aaBBCcDDLL 55	aaBbCcDDLL 52
abCDL	AaBbCCDDLL 58	AaBbCcDDLL 58	AabbCCDDLL 55	AabbCcDDLL 52	aaBbCCDDLL 55	aabbCCDDLL 52	aaBbCcDDLL 52	aabbCcDDLL 49
aBcDL	AaBBCcDDLL 58	AaBBccDDLL 55	AaBbCcDDLL 55	AaBbccDDLL 52	aaBBCcDDLL 55	aaBbCcDDLL 52	aaBBccDDLL 52	aaBbccDDLL 49
abcDL	AaBbCcDDLL	AaBbccDDLL	AabbCcDDLL	AabbccDDLL				

Complete the last row of the Punnett square. **(4 marks)**

Epigenetics

1 State and explain the meaning of the word epigenetic.

....

The literal meaning of epigenetic is 'above genes'.

....

....

.... **(3 marks)**

2 Which base can be methylated in the DNA of mammals?

☐ **A** adenine

☐ **B** thymine

☐ **C** cytosine

☐ **D** guanine **(1 mark)**

3 Explain why evidence that suggests cells can pass on epigenetic memory to their daughters does **not** prove that it can be passed across generations.

Daughter cells can be made in both meiosis and mitosis.

....

....

....

.... **(3 marks)**

4 Describe the structure and function of histones.

....

....

....

.... **(3 marks)**

5 Describe **three** differences between epigenetic changes and genetic mutations of DNA.

....

....

....

.... **(3 marks)**

Biodiversity

1   In a study of birds in Borneo, species abundance was compared between two forest management strategies: sparing and sharing. The sparing strategy involves intensive logging of some plots and leaving others unlogged. The sharing strategy involves light logging of all plots. The data are shown below. Analyse these data to comment on the relative effectiveness of these two strategies for biodiversity conservation.

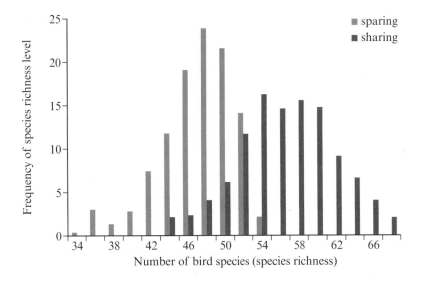

■ sparing
■ sharing

> The word 'analyse' means that you need to comment on the graph **and** then relate your comments to the situation being discussed to make a judgement on the issue.

....

....

....   **(3 marks)**

Practical skills

2   *Pasture on limestone is a habitat with high biodiversity. The grazing of cattle and sheep suppresses the growth of scrub allowing light penetration which maintains a high diversity of small plants. Recently, reduced grazing has occurred in some places and this has allowed brambles and hawthorn to grow, leading to a reduction in biodiversity. Two possible management strategies have been proposed to conserve these habitats: mowing or controlled sheep grazing (conservation grazing). Proponents of conservation grazing believe that sheep grazing is the better method. Design an investigation to test this suggestion.

> Explain how you would carry out this investigation and how you would record and analyse the results. There is an equation for this on page 57 of the Revision Guide. Also make sure your answer is logically structured and give relevant evidence to support your points.

....

....

....

....

....

....

....

....   **(6 marks)**

Adaptation to niches

1 Wood ants are social insects which cooperate with each other in many ways, including finding and gathering food and building nests. The table shows two possible adaptations of wood ants. Complete the table by putting ticks in the boxes to show whether an adaptation is behavioural, anatomical or physiological.

Adaptation	Behavioural	Physiological	Anatomical
production of formic acid as an alarm signal			
acting in a group to carry heavy prey to the nest			

(2 marks)   An adaptation can have more than one origin.

Guided 2 A disinfectant company claims that their product kills 99% of germs. Explain why this claim may **not** be reassuring to users of the product.

It is likely that the 1% which are not killed by the product ....

....

....

.... (3 marks)

3 The development of antibiotic-resistant bacteria is a major concern in the treatment of diseases. When GPs prescribe antibiotics for a bacterial infection they give the advice that patients should complete the course. Explain why this advice is given.

....

....

.... (3 marks)

4 The table shows some information about red and grey squirrels. Red squirrels used to live in woodland throughout the UK. The grey was introduced from the USA and now occupies many parts of the country. Oak is largely a lowland tree whereas pine occurs mainly on higher ground. Explain how the likely distribution of red and grey squirrels in the UK in the present day arose.

	Red squirrel	Grey squirrel
size	1×	2×
food	tree seeds, oak and pine	tree seeds, oak and pine
ability to survive on oak	worse than grey	better than red
ability to survive on pine	good	poor because the seeds are too small

....

....

....

.... (3 marks)

Evolution and speciation

 Maths skills

1 In antirrhinums, flower colour is controlled by a single gene with two alleles that show incomplete dominance. Genotype WW is white, RR is red and RW is pink.

In a study, the frequency of R in a population was found to be 0.1.

(a) Calculate the percentage of flowers that are pink, assuming that the population is in Hardy-Weinberg equilibrium.

> The Hardy-Weinberg equation is $p^2 + 2pq + q^2 = 1$, but it also vital to remember that $p + q = 1$.

Answer .... **(3 marks)**

(b) A few years after this study the population was studied again. The percentage of pink flowers was found to be 32% and that of red was 4%. Analyse this information to explain what may have happened in the intervening years to cause this change.

....

....

....

....

....

.... **(4 marks)**

Guided

2 *Rhododendrons are shrubby plants which are widely distributed throughout the northern hemisphere. The flowering periods and habitats of two species of rhododendron found on Yakushima Island, Japan, are shown in the table. Explain how the two different species of rhododendron may have evolved from a single population of an ancestral species.

> In starred questions (*), structure your answer logically showing how the points you make are related to or follow on from each other. You need to select and apply relevant knowledge of biological facts or concepts to support the argument.

Species	Flowering period	Main flowering period	Habitat
Rhododendron eriocarpum	April to July	May	rocky areas in lowland regions
Rhododendron indicum	May to July	June	high mountainous regions

The original population may have spread into a wider diversity of

habitats ....

....

....

....

.... **(6 marks)**

The classification of living things

1 A new organism has been discovered in a research project exploring a tropical rainforest. The table shows some of its characteristics. Using this information, explain to which of the three domains, Archaea, Bacteria or Eukaryota, the organism belongs.

> Note that there are **three** marks available and **three** features that are unique to the correct domain.

Characteristics	
mitochondria	absent
cell wall containing peptidoglycan	yes
amino acid carried on tRNA that starts protein synthesis	formylmethionine
sensitive to antibiotics	yes
may contain chlorophyll	yes

....

....

....

....   **(3 marks)**

2 *At a recent scientific conference there was a debate as to whether two beetles which had identical morphology but occurred in two very different habitats were actually two different species. Explain how you would investigate this issue by using the reproductive species concept.

> You need to be clear about how to show these organisms can interbreed. This includes checking that they are fertile and that their offspring are fertile.

> In starred questions (*), structure your answer logically showing how the points you make are related to or follow on from each other. You need to select and apply relevant knowledge of biological facts or concepts to support the argument.

....

....

....

....

....

....

....   **(6 marks)**

The validation of scientific ideas

1 Critical evaluation of new data has recently led to new taxonomic groupings.
   From which of the following were these new data derived?

   ☐ **A**   Molecular structure

   ☐ **B**   Molecular phylogeny

   ☐ **C**   Molecular weights

   ☐ **D**   Molecular biochemistry

2 Explain what is meant by the term peer review.

   ....

   > Explain what a review involves and who a peer is.

   ....

   ....

   ....

   ....

   ....   **(4 marks)**

3 Until quite recently, most scientists accepted that all life was either prokaryotic
   or eukaryotic. However, in 1977 Carl Woese published a paper in a scientific
   journal in which he reported the use of molecular phylogenic evidence to suggest
   a third group, the Archaea. Publishing a paper is just one way of informing the
   scientific community of a discovery. State **three** other ways in which he could
   have done this.

   ....

   ....

   ....

   ....   **(3 marks)**

4 Give **two** pieces of evidence from our modern understanding of the
   genetic material that supports Darwin's theory of evolution.

   > This question is directing you to knowledge of DNA that Darwin did not have.

   ....

   ....

   ....   **(2 marks)**

Plant cells

1 A student studied three different cells: an animal cell, a photosynthetic cell from a plant leaf and a non-photosynthetic cell from a plant root.

   (a) DNA is located in the nucleus in

   ☐ **A**   the animal cell only.

   ☐ **B**   the non-photosynthetic cell only.

   ☐ **C**   two of the cells only.

   ☐ **D**   all three cells.   **(1 mark)**

   (b) A cell wall is present in

   ☐ **A**   the animal cell only.

   ☐ **B**   the photosynthetic cell only.

   ☐ **C**   the plant cells only.

   ☐ **D**   all three cells.   **(1 mark)**

   (c) Centrioles are present in

   ☐ **A**   the animal cell only.

   ☐ **B**   the photosynthetic cell only.

   ☐ **C**   two of the cells only.

   ☐ **D**   all three cells.   **(1 mark)**

   (d) A cell surface membrane is found in

   ☐ **A**   the photosynthetic cell only.

   ☐ **B**   the non-photosynthetic cell only.

   ☐ **C**   two of the cells only.

   ☐ **D**   all three cells.   **(1 mark)**

   (e) Amyloplasts may be present in

   ☐ **A**   the photosynthetic cell only.

   ☐ **B**   the animal cell only.

   ☐ **C**   the plant cells.

   ☐ **D**   all three cells.   **(1 mark)**

2 State the names of the structures that allow communication between plant cells.

   ....   **(1 mark)**

Cellulose and cell walls

1 (a) Compare and contrast the structure of a cellulose
molecule with the structure of starch.

> You must give at least **one**
> similarity and **one** difference in
> 'compare and contrast' questions.

....

....

....

....

....

....   **(4 marks)**

(b) Cellulose molecules form cellulose microfibrils.
Explain how the arrangement of cellulose microfibrils
contributes to the physical properties of plant fibres.

> Structure your answer; for example,
> deal with the physical properties
> and then what they contribute.

....

....

....   **(2 marks)**

2 Explain why hydrogen bonds are important in the structure of cellulose.

....

....

....

....   **(3 marks)**

3 Explain how lignin adds strength to xylem tissue.

....

....

....

....   **(3 marks)**

Transport and support

1   Fibre cells in coir have a similar structure to sclerenchyma fibres. Explain how the structure of the coir fibres makes them light, waterproof and strong.

....

....

....

....   **(3 marks)**

2   Answer the following questions about the photograph of some cells in a *Cucurbita* stem.

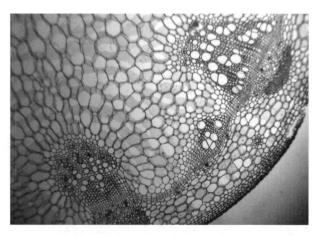

(a) Name the kind of section that has been cut to reveal these cells.

....   **(1 mark)**

(b) Name the cells with rings or spirals in their walls.

....   **(1 mark)**

(c) Name the substance from which these rings and spirals are mainly made.

....   **(1 mark)**

3   The photograph shows a section through a sunflower stem. Draw a low-power plan of it. Label a vascular bundle, xylem, cambium, phloem and sclerenchyma.

In a low-power plan do **not** draw any cells.

(4 marks)

Looking at plant fibres

Practical skills

1 A student determined the mean tensile strength of white and brown coir fibres from the husk of coconut.

Type of fibre	Length of fibre/mm	Tensile strength/MPa ± 2 × SD
white coir	5	192 ± 37
white coir	35	162 ± 32
brown coir	5	343 ± 36
brown coir	35	186 ± 55

(a) Comment on the conclusion that brown fibres are stronger than white.

> Remember that 2 × SD covers 95% of all likely values.

....

....

....

.... **(3 marks)**

(b) Explain how these results could have been obtained.

> This is an 'explain' question, so it is not enough to state your point – you must also say why you have come to that conclusion.

....

....

....

....

....

.... **(4 marks)**

2 Explain why bioplastic is described as a more sustainable form of packaging than oil-based plastics.

....

....

....

.... **(3 marks)**

Water and minerals

1 *In hydroponics, soil is replaced with solutions containing the necessary minerals for plant growth. The mineral solutions are made up with the optimum concentration of each mineral which has been determined by experiment. Design an experiment to determine the optimum concentration of calcium that should be used in the mineral solution used in the growth of a named plant.

> This is a starred (*) question so you need to order your material logically and aim to include scientific terminology and biological evidence where relevant.

....

....

....

....

....

....

....

....

....   **(6 marks)**

2 Complete the table about the features of water that are important to living things.

Feature	Example of importance to living things
high specific heat capacity	
polar solvent	
high surface tension	
incompressibility	
maximum density at 4 °C	

(5 marks)

Developing drugs from plants

Guided

1 In 1775, William Withering tested foxglove extracts to treat the symptoms of a variety of diseases. He used extracts of various parts of this plant to treat 163 patients. Compare the reliability of William Withering's trials with those carried out using contemporary drug testing protocols.

> Look at both similarities and differences.

Withering's study was less reliable because ....

....

....

....

.... **(3 marks)**

2 During the testing of a new antibacterial drug, a double blind trial may be used.

Explain why a double blind trial is important.

....

....

....

....

.... **(3 marks)**

3 Before a new drug can become available for use it has to pass a contemporary drug testing protocol. This includes three-phased testing.

A drug may fail at any of the three phases.

Place a cross (✗) in the box next to the phase at which the drug would have failed. **(2 marks)**

(i) The drug did not improve the condition it was designed to treat in humans.

   ☐ **A** Phase 1

   ☐ **B** Phase 2

   ☐ **C** Phase 3

   ☐ **D** It would not fail

(ii) The effect of the drug was different in humans from its effect in animals.

   ☐ **A** Phase 1

   ☐ **B** Phase 2

   ☐ **C** Phase 3

   ☐ **D** It would not fail

Investigating antimicrobial properties of plants

1   The antimicrobial properties of the extracts of four fruits, apple, guava, orange and pomegranate, were investigated. Cultures of three different bacterial species, A, B and C, were mixed with agar in separate Petri dishes. Small wells were cut into the agar and a different fruit extract was added to each. The Petri dishes were then incubated for 24 hours. After incubation, there were clear zones around each well. The diameter of each clear zone was measured and the results are shown in the graph.

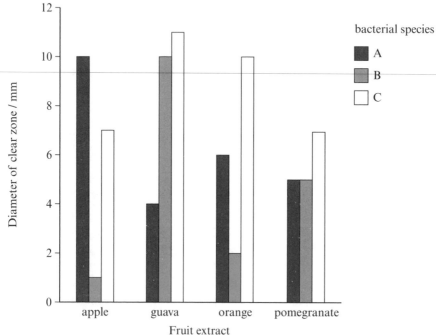

Analyse this information to compare and contrast the effects of the fruit extracts on the different species of bacteria.

....

....

....

....   **(4 marks)**

2   The Jatobá plant, *Hymenaea courbaril*, was tested for antimicrobial properties. Describe how the antimicrobial properties of the seeds of the Jatobá plant could be tested.

....

....

....

....

....   **(5 marks)**

Conservation: zoos

Guided

1 Some tiger species are threatened with extinction in the wild. Captive breeding programmes in zoos are trying to help with this problem. Two new tigers are introduced to the breeding stock every seven years to maintain 90% genetic diversity. Explain why it is necessary to maintain 90% genetic diversity to eventually allow reintroduction of the tigers into the wild.

90% genetic diversity keeps many alleles in the population, thus

.... variety of .... The population needs to show a lot of

phenotypic variation as the tiger lives in a wide range of .... in

the wild. Also, if there were a change in the environment the tigers

would be unlikely to be able .... if genetic variety was ....   **(4 marks)**

2 State **three** concerns about keeping animals in zoos.

....

....

....

....

(3 marks)

> Be careful here **not** to give only animal welfare answers; try to include some scientific concerns as well.

3 The pedigree (family tree) of every racehorse in the UK can be traced back to three male horses imported in the early 1700s. Strict rules that control breeding programmes ensure that every racehorse has its parents listed in a stud book. Describe the likely effect of these breeding programmes on the genetic diversity of racehorses.

> For 'describe' questions, you do not need to make a judgement or explain how or why, you only need to describe what is requested.

....

....

....

....

....

....

(4 marks)

Conservation: seed banks

1   The first withdrawals of seeds from the Svalbard Global Seed Vault were made
    in 2015. These were seeds of ancient varieties of wheat from which our modern
    forms were derived. They were needed to set up new research facilities in
    Lebanon and Morocco to replace the one, which is currently unusable, in Syria.
    Explain why research is being carried out on ancient strains of wheat.

    ....

    ....

    ....

    ....

    ....

    ....   **(3 marks)**

2   *Explain the advantages of conserving plants by using seed banks.

    ....

    ....

    ....

    ....

    ....

    ....

    ....

    ....

    ....   **(6 marks)**

> Make sure you
> structure your answer
> logically, showing how
> the points you make
> are related or follow
> on from each other
> where appropriate.
> You should also
> support your points
> with relevant
> biological facts and/or
> evidence.

3   Explain **two** reasons why it might be considered sensible to conserve a rare plant
    by storing its seeds in a seed bank.

    ....

    ....

    ....

    ....   **(3 marks)**

Exam skills

A study in Zimbabwe showed that areas with elephants have a higher biodiversity than those without. However, further work showed that a density of over 0.5 elephants km^{-2} leads to a reduction in biodiversity. The area studied was $66\,000\,km^2$.

(a) (i) Calculate the elephant population in this area that would achieve maximum biodiversity.

Answer: ....   **(2 marks)**

(ii) The chart shows elephant populations in this area over a 20 year period. An international law giving full protection to elephants was implemented in 1989

*Deduce the likely changes in biodiversity over this period based on your analysis of the data given.

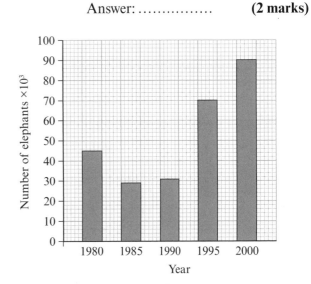

Hint, the * indicates that in your answer to this question the points that you make should be related and follow on from each other. It is a good idea to plan your answer with this in mind. You are also asked to based your answer on your analysis of the data given. Make sure that you put some quantitative analysis into your answer to justify your statements.

....

....

....

....

....

....   **(6 marks)**

(iii) Explain how an index biodiversity could have been measured in each year of the study.

....

....

....

....

....   **(4 marks)**

Ecosystem ecology

1 A student set up a model ecosystem in an aquarium. He put salt water, liquid fertiliser, sand and shell, microbial culture containing more than one type of microbe, and brine shrimps into a clean clear plastic bottle. Evaluate how far this model shows all the features of an ecosystem. Use this information and your own knowledge of the features of an ecosystem.

> In this kind of question where you are given a scenario, you must answer in the context of that scenario.

....

....

....

....

....

.... **(4 marks)**

> 'Evaluate' questions require you to identify the relevant information, showing how the points you make are related, look at the merits and faults and support your judgements by giving evidence.

2 Explain the relationship between populations and communities within an ecosystem.

....

....

....

....

....

.... **(4 marks)**

3 The distribution of plants in a forest is affected by many abiotic factors. Name **one** of these factors and suggest how this factor could affect the distribution of the low-growing plants within the forest.

....

....

.... **(2 marks)**

Distribution and abundance

1   Glaciers are masses of ice that formed thousands of years ago. As a result of warmer temperatures, more ice is melting. This is reducing the length of the glaciers, uncovering bare rock. The diagram shows the length of a glacier 100 years ago and at present. It also shows what is now found in a transect taken from the front edge of the glacier at present, comparing it to where it was 100 years ago. *Epilobium latifolium* is an early coloniser of such bare rock.

Length of glacier 100 years ago					front edge of glacier
Length of glacier at present	bare rock	algae and lichens	mosses	grasses	shrubs and trees

front edge of glacier at present

front edge of glacier

transect line **at** front edge ✗

transect line **from** front edge ✓

(a) Describe how to carry out a study of the distribution of *E. latifolium* from the front edge of this glacier.

> It is always important to read questions very carefully. Here, the word 'from' is important; it means something very different from the word 'at' in this context.

....

....

....

....

....   **(4 marks)**

(b) Explain how you would investigate the possibility of correlation between the distribution of *E. latifolium* and light intensity.

> Measurements will need to be taken of light intensity in an appropriate place in relation to the *E. latifolium* plants.

....

....

....

....   **(4 marks)**

2   Which of the following would be the appropriate sampling method to compare the abundance of a flowering plant in a grazed field and an adjacent ungrazed field?

☐ **A**   controlled

☐ **B**   random

☐ **C**   systematic

☐ **D**   trial and error   **(1 mark)**

Exam skills

1   The graph shows changes in biomass of trees at a site that has been undergoing primary succession for 240 years. The biomass is measured in tonnes per hectare (tonnes ha^{-1}). Analyse the information in the graph to explain the meaning of the term succession.

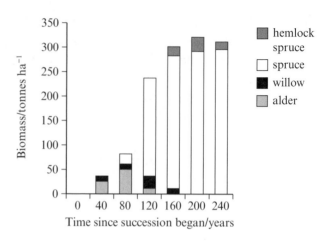

> Make sure your answer combines your knowledge of succession with the context of the information presented.

....

....

....

....

....

.... **(4 marks)**

 Maths skills

2   The diagram shows energy flow measured during a study of a grassland ecosystem. The values shown are kJ m^{-2} year^{-1} × 10^4. Values of energy lost through respiration (R) and other means (L) a re shown for some organisms. The biomass of the organisms in the ecosystem remained unchanged during the study.

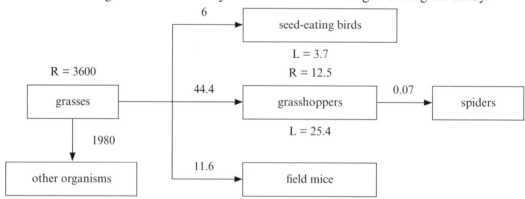

(a)   Calculate the gross primary productivity (GPP) for the grasses in this system. Show your working and the units.

Answer: ....   **(3 marks)**

(b)   Calculate the percentage of the energy taken in by the grasshoppers that is converted into new grasshopper biomass. Show your working.

Answer: ....   **(3 marks)**

Succession

1   The photographs show an area before and after the eruption of Mount St Helens in 1980 in the USA. The graph shows the numbers of plant species before and after the eruption of Mount St Helens. Analyse the information in the photographs and the graph to explain the changes in the number of different plant species in the Mount St Helens area.

Mount St Helens before 1980

Surrounding area immediately after the eruption

Surrounding area in 1994

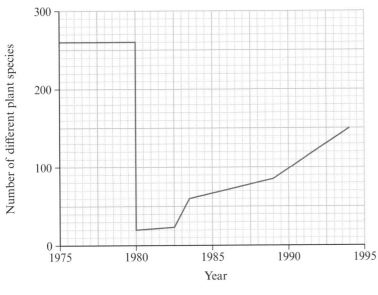

This question is about recolonisation after all the plants had been destroyed by the volcano. Make reference to the shape of the graph **and** what you can see in the photographs. You are asked about the **change** in number of **different types** of plants; this must be the focus of your answer and you must refer to the volcanic effects, for example, the production of ash.

....

....

....

....   **(4 marks)**

>Guided>   2   Describe what is meant by the term climax community.

This is a straightforward factual question but note that there are **three** marks.

A climax community is the final ....

....

....

....   **(3 marks)**

Productivity in an ecosystem

 Maths skills

1 The pie chart shows the distribution of GPP in the Sylt Rømø Wadden Sea, Germany.

The total GPP for this sea is 840×10^6 kJ m^{-2} year^{-1}. Using this information, calculate the GPP for the phytoplankton.

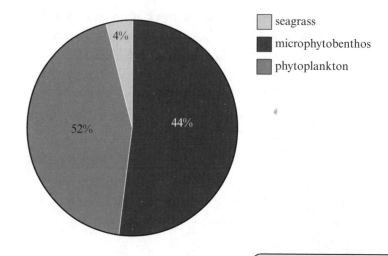

☐ seagrass
■ microphytobenthos
■ phytoplankton

4%

52%   44%

.... (2 marks)

> Do **not** forget the units.

Guided

2 The graphs show the relationship between net primary production and two abiotic factors. Analyse the information in the graphs to explain the relationship between NPP and each of these **two** environmental factors.

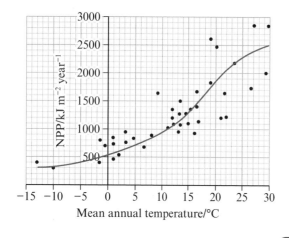

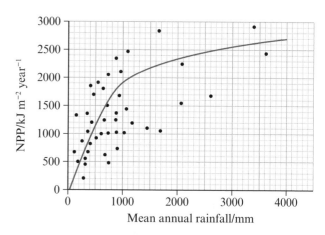

> In a scatter-type graph, especially if a line has been drawn, do not look for ups and downs but for the major trends.

The higher the temperature the more ....

....

....

....

....

....

....

.... (5 marks)

Energy flow

1 The diagram shows the energy flow in a rainforest. Analyse the data to compare the efficiency of energy transfer between the Sun and producers with that between producers and primary consumers.

> The word 'analyse' means that you need to comment on the data and then relate your comments to the situation being discussed to make a judgement.

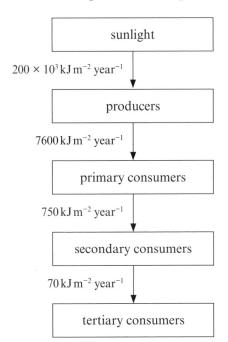

sunlight

$200 \times 10^3 \, kJ \, m^{-2} \, year^{-1}$

producers

$7600 \, kJ \, m^{-2} \, year^{-1}$

primary consumers

$750 \, kJ \, m^{-2} \, year^{-1}$

secondary consumers

$70 \, kJ \, m^{-2} \, year^{-1}$

tertiary consumers

....

....

....

....

....

....   **(4 marks)**

 Maths skills

2 The table shows the fresh biomass of green plants and consumers on an area of grassland. Calculate the efficiency of biomass transfer between the trophic levels.

> This is worked out by looking at the biomass which has been transferred and dividing it by the biomass available in the previous trophic level. Don't forget to multiply by 100 to get a percentage.

Organism	Fresh biomass/g
green plants	2250.0
primary consumers	240.0
secondary consumers	38.0

Answer: ....   **(2 marks)**

Photosynthesis: an overview

1   The diagram shows some of the steps involved in photosynthesis. Complete the
    diagram by writing the correct word or words on the dotted lines.

> Look at the three different pathways separately. Water is made of the elements hydrogen and oxygen;
> which one is used in the light-dependent reaction and which is given out? When carbon dioxide is used
> what is the first organic compound produced? If ATP is converted into ADP what compound is removed?

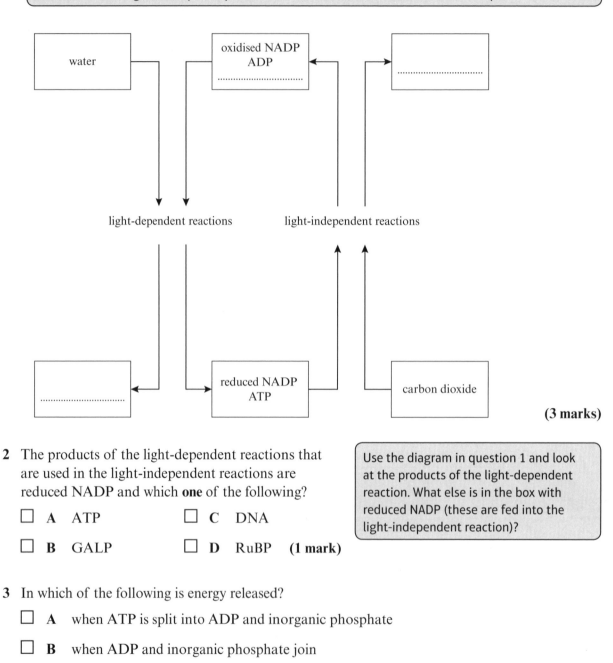

(3 marks)

2   The products of the light-dependent reactions that
    are used in the light-independent reactions are
    reduced NADP and which **one** of the following?

> Use the diagram in question 1 and look
> at the products of the light-dependent
> reaction. What else is in the box with
> reduced NADP (these are fed into the
> light-independent reaction)?

☐ **A**   ATP   ☐ **C**   DNA

☐ **B**   GALP   ☐ **D**   RuBP   **(1 mark)**

3   In which of the following is energy released?

☐ **A**   when ATP is split into ADP and inorganic phosphate

☐ **B**   when ADP and inorganic phosphate join

☐ **C**   when inorganic phosphate forms bonds with water

☐ **D**   when inorganic phosphate and water are separated   **(1 mark)**

The light-dependent reactions

1 Explain how oxygen is produced during the light-dependent reactions of photosynthesis.

Do **not** just write all you know about the LDRs; you also need to explain **how** oxygen is produced.

....

....

....

....   **(3 marks)**

2 The equation shows a reaction in photosynthesis:
water → hydrogen ions + oxygen gas + electrons

Describe the role of the electrons produced in this reaction.

Again, answer the question which has 'in this reaction' in it.

....

....

....

....

....   **(4 marks)**

3 (a) When light is absorbed by chlorophyll, it excites

☐ **A** electrons

☐ **B** neutrons

☐ **C** photons

☐ **D** protons.   **(1 mark)**

(b) Oxygen is produced when water molecules are split in the process of

☐ **A** analysis

☐ **B** autolysis

☐ **C** hydrolysis

☐ **D** photolysis.   **(1 mark)**

The light-independent reactions

1 Oxygen inhibits the enzyme that catalyses the fixing of carbon dioxide. High concentrations of oxygen within a chloroplast can reduce the rate of photosynthesis. Explain the effect of high concentrations of oxygen on the rate of carbohydrate production in a chloroplast.

> As in many of the questions, you have to combine information you have been given with your own knowledge.

....

....

....

....

....

....   **(4 marks)**

2 In the Calvin cycle, six RuBP molecules are made from six turns. Twelve molecules of GP are made.

(a) Calculate the difference in the number of carbon atoms in these two sets of molecules.

Answer: ....   **(2 marks)**

(b) Explain the difference you found in part (a).

....

....   **(2 marks)**

3 At which stage does RuBP combine with carbon dioxide?

☐ **A** the light-dependent reactions of the Calvin cycle

☐ **B** the light-independent reactions of the Calvin cycle

☐ **C** the light-dependent reactions of the Krebs cycle

☐ **D** the light-independent reactions of the Krebs cycle   **(1 mark)**

Exam skills

1   A transect which extended inland from a beach was used to find the distribution of plant species on sand dunes. Quadrats were used at nine positions along the transect. The percentage cover and number of plant species in each quadrat were recorded. A soil sample from each quadrat was used to measure the mass of organic material present. The results are shown.

Quadrat number	1	2	3	4	5	6	7	8	9
bare sand (% cover)	80	30	30	8					
sea couch (% cover)	20								
marram grass (% cover)		70	50	20	5	5			
red fescue (% cover)			5	40	55	40			
sea buckthorn (% cover)							80		
common heather (% cover)								90	
Corsica pine (% cover)									100
distance from top of beach/metres	0	80	170	250	500	650	980	1600	1980
number of species found	1	1	5	11	18	7	5	2	2
mass of organic material/grams	0.4	0.3	0.3	0.9	2.8	6.4	25.1	23.4	32.8

(a)   *Analyse this information to evaluate the conclusion that the results show evidence of succession.

> You must make reference to the data given and not just write about succession. Also, make sure your answer is logically structured, showing how your points are linked, and select and apply relevant facts or concepts to support your points.

....

....

....

....

....

....

....

....

....

....   **(9 marks)**

(b)   Explain how you could show that a named species is absent from certain sample points due to competition and **not** the soil conditions present at each sample point.

....

....

....   **(3 marks)**

Chloroplast

1 The diagram shows a plant cell. In which
  structure would DCPIP be reduced
  during the Hill reaction?

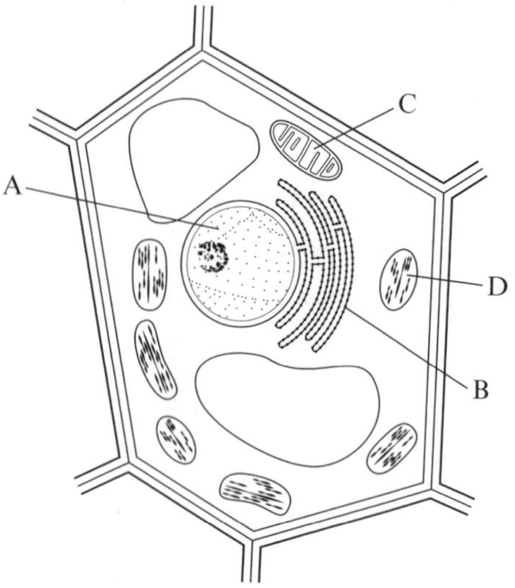

  ☐ A

  ☐ B

  ☐ C

  ☐ D   (1 mark)

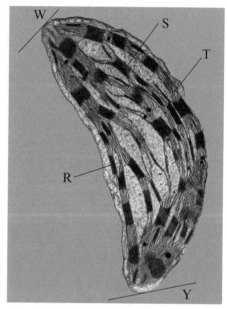

 Maths skills

2 The photograph shows a chloroplast.

  (a) Calculate the actual length of this chloroplast,
      between the lines labelled W and Y.

  > You will need to use and
  > manipulate the magnification
  > formula to answer this
  > question: magnification = size
  > of image ÷ size of real object.

          Answer: ....   **(3 marks)**   Magnification × 7500

  (b) State the name and function of the structures labelled R, S
      and T.

  > This is a synoptic question.
  > It combines material from
  > topic 4 and topic 5.

  ....

  ....

  ....

  ....   **(3 marks)**

Climate change

1 The table shows some data from pollen analysis of a peat bog in Finland. The diagram shows where four tree species found in the pollen study are distributed today with respect to climate zone.

Depth of sample/m	Age/years	Tree pollen grain (%)			
		Larch	Spruce	Pine	Beech
0.5	2850	0	0	53	43
1.0	3770	0	0	55	40
1.5	5600	0	0	31	47
2.0	6390	0	12	15	53
2.5	8170	5	36	4	48
3.0	8700	38	36	6	35
3.5	8780	27	40	3	32
4.0	10000	10	22	2	40

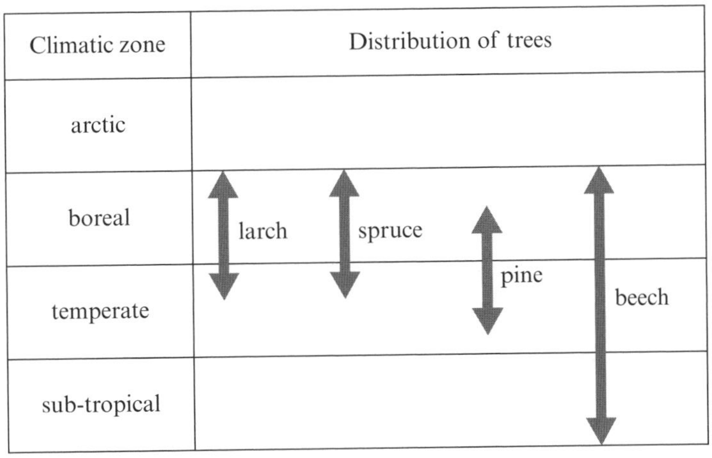

Climatic zone	Distribution of trees
arctic	
boreal	larch spruce pine beech
temperate	
sub-tropical	

*Analyse all the information to explain in what way the climate in Finland has changed during the last 10 000 years.

....
....
....
....
....
....

In starred questions (*), structure your answer logically showing how the points you make are related to or follow on from each other. You need to select and apply relevant knowledge of biological facts or concepts to support the argument.

The 'explain' means you have to describe how the climate has changed and then come up with supporting evidence from the data.

(6 marks)

2 Explain how increases in carbon dioxide and methane, released into the atmosphere, may be contributing towards the estimated changes in mean global temperature.

Both are greenhouse gases; methane is much more powerful but less in amount.

....
....
....

(3 marks)

Had a go ☐ Nearly there ☐ Nailed it! ☐

Anthropogenic climate change

1 'The burning of fuels will lead to global warming.' Criticise this statement.

> 'Criticise' tells you to inspect a set of data, an experimental plan or a scientific statement and consider the elements. Look at the merits and faults of the information presented and support judgements made by giving evidence.

....

....

....

....

....

.... **(4 marks)**

2 The graph shows the changes in mean global surface temperature between the years 1880 and 2000.

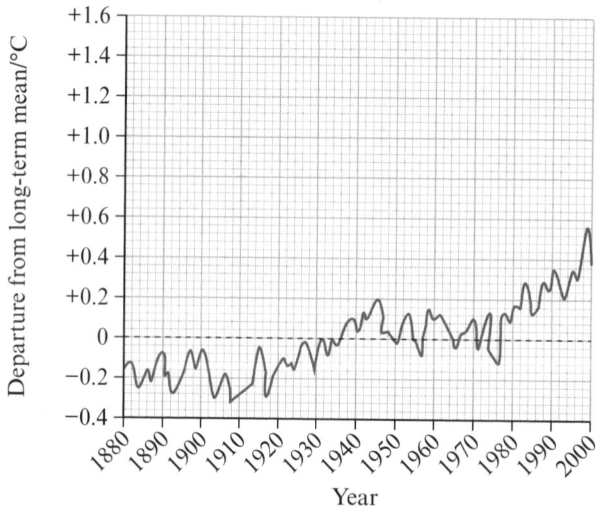

(a) Draw the line of best fit on the graph between 1940 and 2000.

(1 mark)

> In this case, the line will **not** be straight.

(b) Analyse this information to explain why a predicted increase of 0.6 °C in temperature between 2000 and 2020 might **not** be accurate.

....

....

....

....

....

.... **(4 marks)**

The impact of climate change

1 A study was carried out on the rate of oxygen consumption of germinating peas at two temperatures, 10 °C and 20 °C. The graph shows the results and statistically fitted trend lines. Calculate the Q_{10} for oxygen consumption of these peas.

$$Q_{10} = \frac{\text{initial rate at } (T + 10\,°C)}{\text{initial rate at } T\,°C}$$

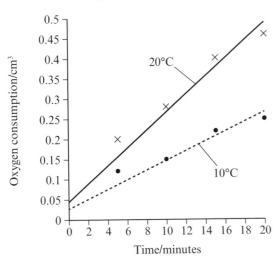

Q_{10} = ... **(3 marks)**

2 The graphs show the changes in midsummer maximum air temperature and the upper limit of two intertidal invertebrates, barnacles and mussels. Explain how far this information supports the suggestion that global warming may lead to changes in the distribution of living organisms.

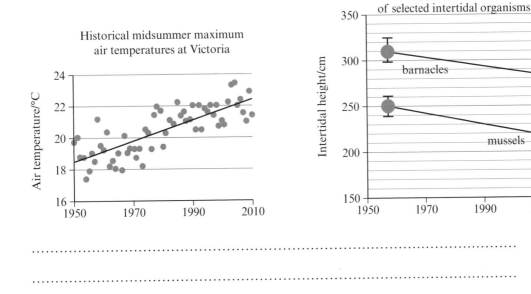

....

....

....

....

....

....   **(4 marks)**

The effect of temperature on living things

Practical skills

1 A student carried out an investigation to determine the effect of temperature on the hatching success of brine shrimp eggs. The results are shown in the table. Analyse the data to explain how global warming, directly or indirectly, might alter brine shrimp populations.

Temperature/°C	22	24	26	28	30	32
Eggs hatching (%)	36	40	42	44	42	41

> Notice the words 'directly' and 'indirectly'.

....

....

....

....   **(3 marks)**

Guided

2 The scatter diagram shows differences in the time of leaf bud opening between oak and ash in relation to environmental temperature in the UK. Each dot represents one year.

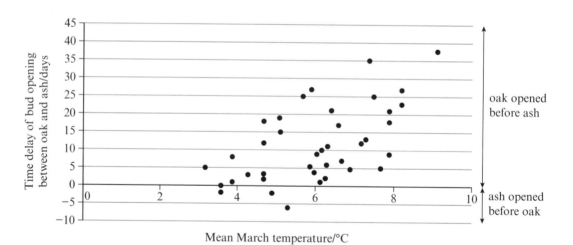

Mean March temperature/°C

(a) Explain how global warming may affect the relative abundance of oak and ash in woodland where both species grow together.

Oak usually opens before ash ....

....

....   **(2 marks)**

(b) Explain how earlier leaf bud opening might affect the abundance of small plants that grow on the woodland floor.

....

....

....   **(2 marks)**

Decisions on climate change

1 Explain what is meant by the word biofuel.

....

.... **(2 marks)**

2 Explain why a forest might be a net absorber of CO_2 (a carbon sink) at one point in time, but a carbon store and not a net absorber at another point in time.

....

....

....

.... **(4 marks)**

3 The graph shows changes in world palm oil production between 1995 and 2004. 85% of this palm oil is produced in Indonesia and Malaysia. Much of the forest area inhabited by orangutans has a suitable climate for the production of palm oil, and the demand for palm oil is increasing for use in food production and biofuels. Orangutans avoid palm oil plantations. Demand is predicted to be double the 2004 value by 2020, as biofuels become more widely used, replacing fossil fuels.

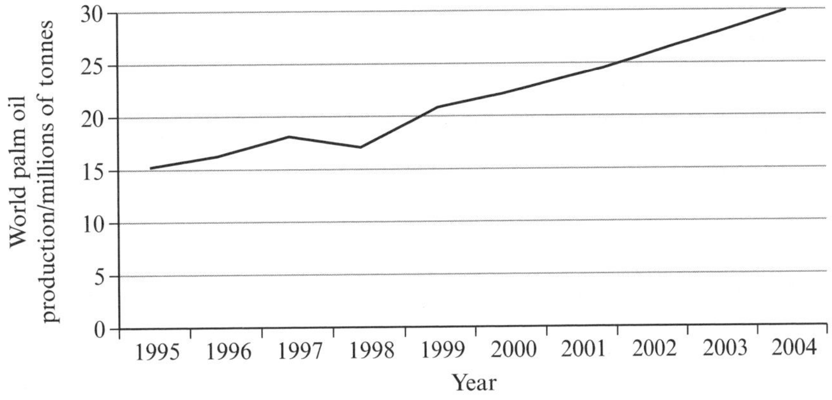

Maths skills
(a) If Indonesia and Malaysia continue to produce 85% of the world supply of palm oil, calculate how many tonnes of palm oil they will produce in 2020 if predictions are correct.

Answer: .... **(3 marks)**

Guided
(b) Discuss the benefits and problems caused by the planting of palms to produce oil. Your answer should refer to the effects on humans, orangutans and the wider environment.

> In a question such as this, you must address both sides of the argument.

Biofuel is less directly polluting ....

....

....

....

.... **(4 marks)**

Evolution by natural selection

1 W, X, Y and Z are related species. Y and Z are evolved from species X. Species W is the least related to the others. Which of the phylogenetic trees below represents these relationships?

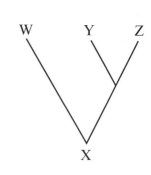

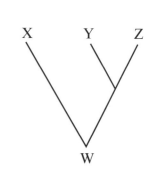

Your answer ....   **(1 mark)**

2 Which of the following is an assumption of the Hardy-Weinberg equilibrium?

☐ **A** Mating must be non-random.

☐ **B** The population must be large.

☐ **C** There should be movement of organisms into and out of the population.

☐ **D** There must be mutations.   **(1 mark)**

3 Which of the following may lead to a directional change in allele frequency?

☐ **A** non-random mating   ☐ **C** mutation

☐ **B** a small population   ☐ **D** selection pressure   **(1 mark)**

4 State **one** piece of evidence that shows that all life had a common ancestor.

.... **(1 mark)**

Speciation

1 *The hawthorn maggot fly (*Rhagoletis pomonella*) of the USA feeds on hawthorn. However, since the cultivation of apple orchards began in the USA in the mid-eighteenth century, the fly has been able to use apples as a food source. Laboratory studies have shown that those flies completing their life cycle on apple (the apple race) do so in under 40 days, whereas those on hawthorn (the hawthorn race) take over 50 days. In the wild, the apple race of flies emerges about six weeks before the hawthorn race. The flies complete their entire life cycle on the fruit wherever the female first laid her eggs. Analyse the information to explain why this might be regarded as an example of sympatric speciation in progress.

> In starred questions (*), structure your answer logically showing how the points you make are related to or follow on from each other. You need to select and apply relevant knowledge of biological facts or concepts to support the argument.

> You need to define sympatric speciation and then describe this particular fly's life cycle to show how it meets with the definition.

....

....

....

....

....

....

....   **(6 marks)**

2 There are two species of rhinoceros in Africa: the white rhinoceros and the black rhinoceros. The white rhinoceros feeds on grasses. It has a shoulder height of 1.5–1.8 m and has broad flat lips. The black rhinoceros eats the leaves of shrubs. It has a shoulder height of 1.4–1.7 m and has a pointed mouth. Explain how these two species of rhinoceros evolved from their common ancestor.

> In questions like this, where you are given a scenario, make sure you refer to the context provided and do **not** just write a generic answer.

....

....

....

....

....

....

....   **(5 marks)**

Death, decay and decomposition

1 (a) The time of death of a person can be estimated by relating the time of death to the ambient (surrounding) temperature, the core temperature and the mass of the body. Explain why the ambient temperature and the core temperature of the body are used to determine the time of death.

....

....

....

.... **(3 marks)**

>**Guided** (b) The time of death of a naked body lying stretched out and in still air was estimated to be 24 hours ago. Explain how the estimated time of death would change for each of these factors: if the body had been clothed, if it had been curled up or if it was in moving air.

> Each of your statements should have a conclusion about the effect, followed by a reason for that conclusion.

For the clothed body the estimate was too short because ....

....

....

....

....

....

.... **(5 marks)**

2 The table shows when particular groups of organisms are found on a human body after death. A forensic entomologist was asked to determine the time of death of a person found dead in a wood. No clothing or personal belongings were found. Organisms from groups C, D and E were found on the body. The time of death was estimated as nine days previously. Analyse the data to evaluate this estimate of time of death.

> 'Evaluate' means inspect a set of data, an experimental plan or a scientific statement and consider the elements. Look at the merits and faults of the information presented and support judgements made by giving evidence.

Group	Time after death/days									
	1	2	3	4	5	6	7	8	9	10
A	✓	✓	✓							
B			✓	✓	✓	✓				
C					✓	✓	✓	✓	✓	✓
D							✓	✓	✓	
E									✓	✓

....

....

....

.... **(3 marks)**

DNA profiling

1   The diagram shows how substances X, Y and Z are involved in the PCR. It also gives the temperature treatments T1, T2 and T3 at various stages.

| The DNA sample is mixed with X, Y and DNA polymerase. This mixture is placed into a PCR machine. | → | The mixture is treated (T1) to separate Z. | → | The mixture is treated (T2) to allow X to attach to Z. | → | The mixture is treated (T3) to allow DNA polymerase to use Y to assemble a complementary Z. | → | The cycle of heating and cooling is repeated approximately 30 times. |

(a)   Name substances X, Y and Z.

....

....

.... **(3 marks)**

(b)   Match the temperature regime to the letters T1, T2 and T3.

heated to 90–95 °C   ....

heated to 75 °C   ....

cooled to 55–60 °C   ....

cooled to 4 °C   ....

> PCR is required by the specification so you should learn these details. One of the answers is a decoy.

(3 marks)

2   Describe how gel electrophoresis can be used to analyse DNA.

> Notice the starting point here is DNA.

....

....

....

....

....

.... **(4 marks)**

3   Describe how DNA profiles of two closely related species would be compared.

....

.... **(3 marks)**

Exam skills

1   The tolerance of plants to copper ions in the soil is under genetic control. The frequency of an allele which causes a plant to be more tolerant to copper ions was measured at two different sites, A and B. The table shows the percentage frequencies of the tolerance and non-tolerance alleles in plant populations at the two sites.

Site	Percentage frequencies of	
	Tolerance allele	Non-tolerance allele
A	30	70
B	80	20

(a) Explain what is meant by the frequency of an allele in a population.

....

....

.... **(2 marks)**

(b) Describe how natural selection could have brought about the different allele frequencies at the two sites.

> You must be careful to write an answer in the context of the information given and **not** just something generic.

....

....

....

....

....

.... **(4 marks)**

(c) Suggest why bacteria often adapt to changing conditions much more quickly than plants.

....

....

....

.... **(3 marks)**

Bacteria and viruses

Guided 1 Describe the structure of a virus.

> You must be careful **not** to make a point which only relates to a specific virus, such as possessing gp120, unless you base your whole description on that named virus to which the specific point applies.

The genetic material of viruses is in the form of DNA or RNA ....

....

....

....

....

....   **(4 marks)**

2 State **three** ways in which the structure of *Mycobacterium tuberculosis* differs from that of a virus.

....

....

....

....   **(3 marks)**

3 Infection with Human Immunodeficiency Virus (HIV) increases the risk of developing tuberculosis (TB). Tuberculosis is caused by the bacterium *Mycobacterium tuberculosis*. The table shows the results of a survey of patients who had TB in 2008 and 2010. It shows the number of patients with TB who believed that they were HIV negative and the number of patients who knew that they were HIV positive.

Year	Number of patients with TB $\times 10^3$	
	HIV negative	**HIV positive**
2008	600	800
2010	1600	500

Maths skills (a) Calculate the percentage of patients with TB in 2008 who were HIV positive.

> Always make an estimate to check your calculated answer when doing percentage calculations; in this case, you would suggest just over half or 50%.

Answer: ....   **(2 marks)**

(b) Describe how the proportion of patients who were HIV positive in 2008 compares with the proportion of patients who were HIV positive in 2010.

....

....

....   **(2 marks)**

Pathogen entry and non-specific immunity

1 Explain how gut flora protect the body from infection.

....

> Remember that the gut flora and the infectious agents are both microbes.

....

....

.... **(3 marks)**

2 *The diversity of the gut flora of a person was recorded. This person then took a course of antibiotics for seven days. The percentage of each type of bacterium in the gut flora was recorded for a period of 18 months. Each type of bacterium in the graph is represented by a different letter. Analyse the information in the graph to explain the effect of this course of antibiotics on the diversity of gut flora.

> As in many questions, you need to describe (in the 'analyse' part) and then explain what you have described.

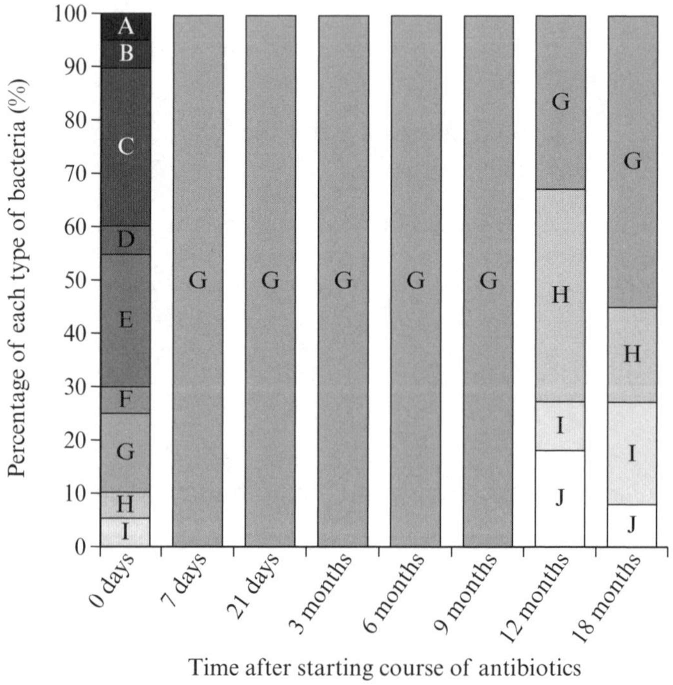

Time after starting course of antibiotics

....

....

....

....

....

....

....

....

....

....

.... **(9 marks)**

Specific immune response – humoral

1 The table below describes some features of the lymphocytes which are involved in the immune system. Place a tick in the appropriate column to indicate whether the description is true or false.

Description	True	False
B and T cells are formed in the bone marrow		
B cells stimulate T cells to produce clones of memory cells		
T helper cells produce chemicals that destroy pathogens		
B and T cells are able to form clones by mitosis		

(4 marks)

2 The antibody 2G12 is produced in response to part of a glycoprotein found on the surface of HIV. Synthetic molecules have been made that resemble this part of the glycoprotein. The antibody 2G12 binds to these synthetic molecules. Explain how this may enable scientists to develop a means of producing active immunity to HIV infection.

....

....

....

....

....

.... (4 marks)

3 Methicillin-resistant *Staphylococcus aureus* (MRSA) is a bacterium. When it enters the blood it can stimulate the production of several different clones of plasma cells. These produce a variety of antibodies (polyclonal antibodies). Suggest an explanation for this.

....

....

....

....

....

.... (4 marks)

Specific immune response – cell-mediated

1   A person who has had an organ transplant has to take immunosuppressive drugs. This prevents the immune system from destroying the organ transplant. Some of these drugs work by inhibiting the production of cytokines. Explain how these drugs might affect a person who has had a transplant if they become infected with a bacterium or a virus.

> Say what happens and why.

....

....

....

....

....

.... **(4 marks)**

2   When the adenovirus infects someone for the first time, an immune response occurs and the person develops immunity. T killer cells are involved in the immune response to the adenovirus.

(a) State the kind of cell that presents antigens to T killer cells.

.... **(1 mark)**

(b) Name the chemical that T helper cells produce to activate the T killer cells.

.... **(1 mark)**

(c) Name the process by which T killer cells divide once activated.

.... **(1 mark)**

▷Guided▷   (d) Describe the role of T killer cells in the immune response to a viral infection.

T killer cells are involved in the destruction of virus-infected host cells

by ....

....

....

....

....

.... **(4 marks)**

Post-transcriptional changes to mRNA

1 The following diagram shows the production of the mRNA from the gene for α-tropomyosin. State how many introns and how many exons this gene has.

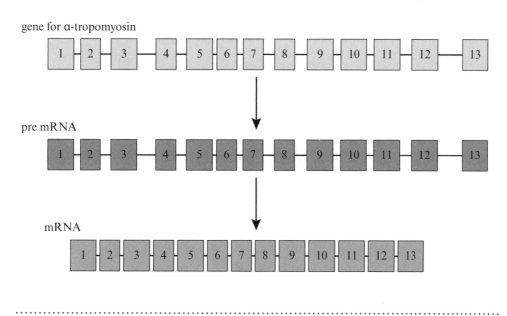

gene for α-tropomyosin

| 1 | 2 | 3 | 4 | 5 | 6 | 7 | 8 | 9 | 10 | 11 | 12 | 13 |

pre mRNA

| 1 | 2 | 3 | 4 | 5 | 6 | 7 | 8 | 9 | 10 | 11 | 12 | 13 |

mRNA

| 1 | 2 | 3 | 4 | 5 | 6 | 7 | 8 | 9 | 10 | 11 | 12 | 13 |

.... **(1 mark)**

2 Explain **two** ways in which one gene could give rise to more than one protein.

> It says 'explain' so you must state a way and then say how it gives rise to more than one protein.

....

....

....

....

....

.... **(4 marks)**

Guided 3 The average gene encodes for about five different forms of the same protein, and there are genes (neurexin B, for example) that encode for over 1000. Explain why this means a significant rethink of the one gene-one enzyme hypothesis of Beadle and Tatum is needed.

Enzymes are made of protein ....

....

....

.... **(3 marks)**

Types of immunity

1 The following list gives examples of how immunity can develop:

   P   Antibodies are transferred into the blood of a baby from its mother before birth.

   Q   Killer T cells are produced by the body when it is infected by a virus.

   R   The polio virus, which has been made incapable of replicating, is given to babies to stimulate the production of memory cells.

   S   Anti-venom contains antibodies produced in an animal. This anti-venom can be injected to give protection against snakebite venom.

Complete the table below by writing a letter from the list above to match the type of immunity described in each case.

Immunity	Active	Passive
natural		
artificial		

              (4 marks)

2 Whooping cough is caused by the bacterium *Bordetella pertussis*. In an investigation, a group of rats was vaccinated against *B. pertussis*. Sixty days later these rats were experimentally infected with *B. pertussis*. The graph shows the mean levels of antibody A and antibody B throughout.

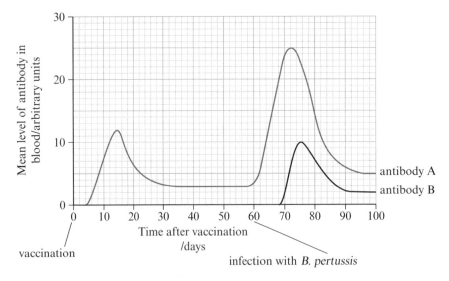

Make sure you pick only the relevant information from the graph; it is **not** all needed for this question.

(a) Explain the changes in mean level of antibody A after infection with *B. pertussis*.

....

....

.... **(3 marks)**

(b) Explain why antibody B was **not** present in the blood of these rats until after infection with *B. pertussis*.

....

.... **(2 marks)**

Antibiotics

Practical skills

1 The antibiotics vancomycin and tetracycline act on bacteria in different ways: vancomycin affects the wall of the cells and tetracycline acts on ribosomes.

(a) State why vancomycin does **not** affect human cells.

.... **(1 mark)**

(b) Explain how tetracycline prevents bacterial growth.

.... **(2 marks)**

2 (a) Cephalosporins are antibiotics that inhibit the production of bacterial cell walls. Explain why cephalosporins are bactericidal antibiotics.

.... **(2 marks)**

(b) Quinolones are antibiotics that inhibit the synthesis of DNA in bacterial cells. Explain why quinolones are bacteriostatic antibiotics.

....

....

.... **(2 marks)**

> These questions are synoptic so you need to think about what you have learned about bacterial structure and about DNA and enzymes to answer them.

Practical skills

3 (a) Practical work in microbiology requires the aseptic transfer of bacteria from a liquid culture to other types of culture. Name the instrument used to transfer bacteria from a liquid culture to an agar plate.

.... **(1 mark)**

(b) Describe how agar plates are prepared and how bacteria are transferred from a broth on to them.

....

....

....

....

....

.... **(4 marks)**

Evolutionary race

 Guided   1   The overuse of antibiotics is causing concern. The table shows the number of prescriptions for antibiotics per 10 000 population in the USA given during treatment for influenza from 2000 to 2006.

Year	Number of prescriptions per 10 000 population
2000	226
2002	164
2004	172
2006	142

Maths skills   (a)   The target set by health authorities in the USA for the number of prescriptions per 10 000 population by 2012 is 128, an overall reduction of 43.4% since 2000. Analyse the data to predict whether this target will be achieved.

> Here, 'analyse' means you have to do some calculations to get to an answer which is supported by the data.

The fall from 2000 to 2006 is 226 − 142 = 84, which is .... %

....

....

....

....

....

....   **(5 marks)**

(b)   Explain why health authorities in the USA are encouraging the reduction in the number of prescriptions of antibiotics.

....

....

....   **(2 marks)**

2   There is an 'evolutionary race' between some bacteria, such as *Mycobacterium tuberculosis* (TB), and their hosts. Suggest how this could affect antigen presentation to T helper cells.

....

....

....

....

....

....   **(4 marks)**

Exam skills

The graphs show changes in the quantities of an antibiotic used in a hospital and the percentage of infections caused by bacteria resistant to the antibiotic over the same time period.

> You need to structure your answer logically and aim to support your points with relevant evidence/facts.

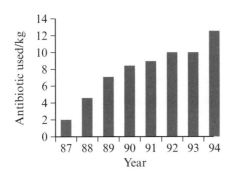
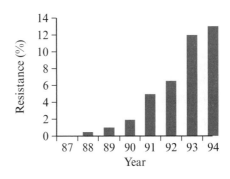

(a) *Analyse the information in the graphs to explain how bacteria become resistant to antibiotics.

> Start your answer by referring to the graphs. This will ensure that you do not forget to do so.

....

....

....

....

....

....

....

....

  (6 marks)

(b) Bacteria were grown on an agar plate and incubated with four different antibiotics. The antibiotics were placed on paper discs. The resulting plate is shown in the diagram. It was concluded that antibiotic B is the most effective. Give reasons why this conclusion may **not** be justified from these data.

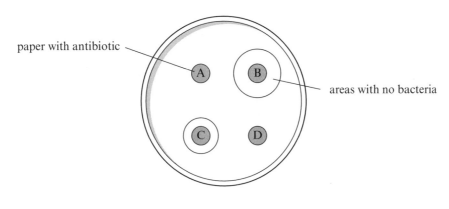

paper with antibiotic

areas with no bacteria

....

....

....   **(3 marks)**

Skeleton and joints

1 Explain why muscles occur in antagonistic pairs.

....

....

....

.... **(3 marks)**

2 Muscles, bones and joints allow movement of the skeleton. Place a cross in the box next to the correct word to complete each of the following statements.

(a) Muscles are attached to bones by

☐ **A** cartilage

☐ **B** ligaments

☐ **C** synapses

☐ **D** tendons.   **(1 mark)**

(b) In a joint, bones are joined to each other by

☐ **A** cartilage

☐ **B** ligaments

☐ **C** synapses

☐ **D** tendons.   **(1 mark)**

3 The diagram shows the arrangement of muscles and bones in an arm. A 5 kg mass was held steady in the position shown and then lifted upwards towards the body. Explain what is happening to muscles A and B and joint D when holding the mass steady and when lifting it.

> Why not try it on your own arm?

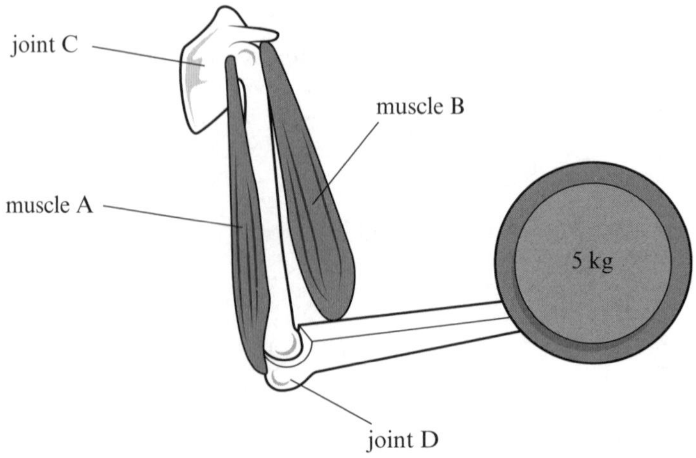

....

....

....

.... **(3 marks)**

Muscles

1   Muscles, tendons and the skeleton all interact when a human leg moves. The diagram shows part of a muscle fibre. Circle the correct answer.

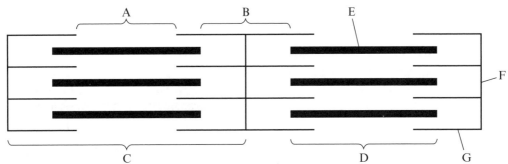

(a)   Which label shows a sarcomere?

☐ **A**   ☐ **B**   ☐ **C**   ☐ **D**   **(1 mark)**

(b)   What happens to the length of the sarcomere when a muscle contracts?

☐ **A**   becomes zero   ☐ **C**   increases

☐ **B**   decreases   ☐ **D**   stays the same   **(1 mark)**

(c)   Which label shows a place where tropomyosin is found?

☐ **A**   ☐ **E**   ☐ **F**   ☐ **G**   **(1 mark)**

(d)   Which ions are released to bind to troponin?

☐ **A**   calcium   ☐ **C**   potassium

☐ **B**   phosphate   ☐ **D**   sodium   **(1 mark)**

(e)   Which is the thin filament in a muscle fibre?

☐ **A**   actin   ☐ **C**   ATPase

☐ **B**   ATP   ☐ **D**   myosin   **(1 mark)**

2   Explain the role of ATP in muscle contraction.

> ATP does what it nearly always does, supplies energy, but for what?

....

....

....

.... **(3 marks)**

The main stages of aerobic respiration

1 The diagram shows some of the stages of aerobic respiration.

(a) Using only this information, calculate the ATP yield from one molecule of glucose.

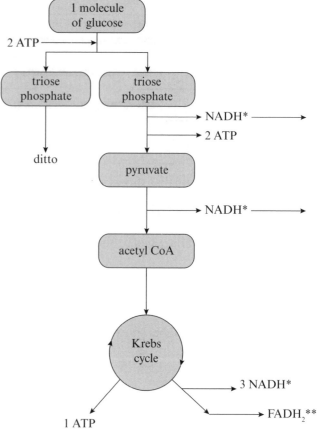

* 1 molecule of NADH yields 3 molecules of ATP in the electron transport chain
** 1 molecule of $FADH_2$ yields 2 molecules of ATP in the electron transport chain

> It is very important to set out calculations like this logically to keep track of what you are doing, for example:
>
> ?TP requires ?ATP to be made so this is ?
>
> ?TP yields ?ATP so ? + ? = ? etc.
>
> (each ? represents a number)

Answer: ....   (3 marks)

(b) Describe how the NADH and $FADH_2$ from glycolysis and the Krebs cycle are involved in the production of ATP.

> They both have basically the same role.

....

....

....

....

....

....

(4 marks)

Glycolysis

1 The diagram shows some
   of the stages in glycolysis
   using the hexose sugar
   glucose.

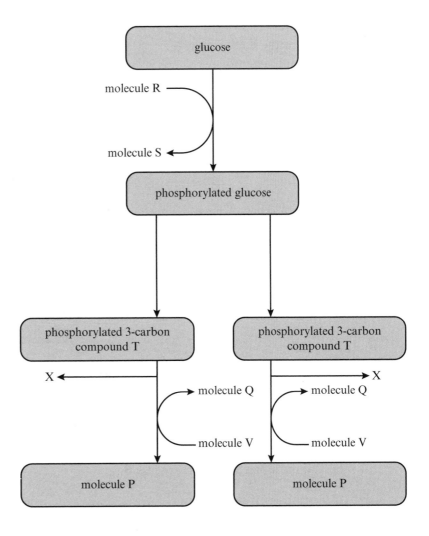

(a) Name R, S, Q, V and X.

R: ....

S: ....

Q: ....

V: ....

X: .... **(5 marks)**

(b) State the number of molecules of R and Q that would be involved.

....

.... **(2 marks)**

(c) Explain why the reaction involving molecules R and S shown
     in the diagram allows the reactions of glycolysis to continue.

> Think of a fire; it needs
> energy to start it but
> then gives energy out.

....

....

.... **(3 marks)**

Link reaction and Krebs cycle

⟩**Guided**⟩ 1 The diagram shows an outline of the Krebs cycle and the link reaction.

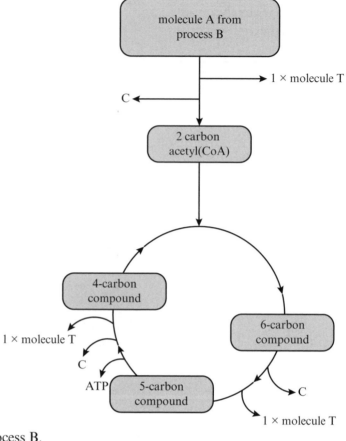

(a) Name molecule A and process B.

Molecule A is pyruvate ....

.... **(2 marks)**

(b) Name molecule T and explain your answer.

....

....

(2 marks)

> Look at the number of molecules in the precursors to each of these reactions.

(c) State what would be produced at C.

.... **(1 mark)**

(d) Explain what would happen in the Krebs cycle if acetyl CoA became unavailable.

....

....

.... **(3 marks)**

Oxidative phosphorylation

1   The diagram shows part of the process of chemiosmosis in a mitochondrion.

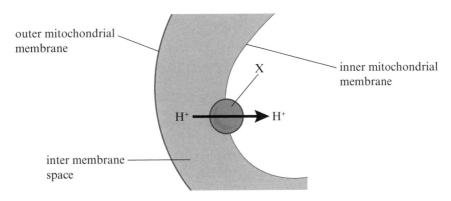

outer mitochondrial
membrane

X

inner mitochondrial
membrane

H⁺ → H⁺

inter membrane
space

(a) Name molecule X.

.... **(1 mark)**

(b) Explain how the concentration gradient of hydrogen ions (H^+) is maintained.

> This is an 'Explain' question, so it is not enough to state your
> point – you must also say why you have come to that conclusion.

....

....

....

....

....

.... **(4 marks)**

(c) Explain what is meant by the term oxidative phosphorylation.

....

....

....

.... **(3 marks)**

Had a go ☐ Nearly there ☐ Nailed it! ☐

Anaerobic respiration

1 Give an account of the anaerobic respiration of glucose in a mammalian muscle cell.

> This is a process so it is important that the events are linked in the correct order.

....

....

....

....

....

....

.... **(5 marks)**

2 Explain why a period of anaerobic respiration leads to an oxygen debt.

> Think about what happens to the lactate.

....

....

....

.... **(3 marks)**

Maths skills

3 The maximum oxygen intake of a runner is 3.5 dm^3 min^{-1}. An oxygen debt of 14.5 dm^3 can be incurred. The runner needs oxygen at a rate of 0.3 dm^3 s^{-1} to run at 5 m s^{-1}. Calculate the maximum distance the runner can cover before being overcome by exhaustion.

> Make sure the oxygen intake values all have the same units.

Maximum distance runner can cover .... **(3 marks)**

The rate of respiration

1 The apparatus shown was used to measure the rate of respiration of germinating seeds in air. The experiment was repeated with the air replaced by nitrogen gas. The rate of respiration of small insects in air was measured using the same apparatus.

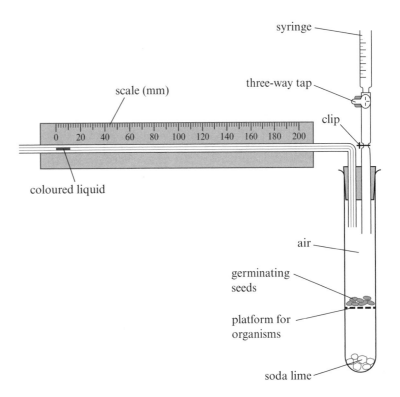

(a) Explain the purpose of the soda lime.

....

.... **(2 marks)**

(b) Explain how the syringe and three-way tap allowed replicate measurements to be made.

....

This is a core practical so you need to learn these details.

.... **(2 marks)**

(c) The table shows the results. Calculate the mean rate of respiration for the insects and complete the table.

Organism	Distance moved by liquid in 15-minute intervals/mm				Mean rate of respiration/ mm min^{-1}
germinating seeds	7	6	5	6	0.4
germinating seeds in nitrogen gas	0	0	0	0	0
insects	12	11	13	12	

First you need to find the mean distance moved in the insect experiment.

Mean distance moved by liquid in 15 minutes for insect experiment = ... **(2 marks)**

(d) The seeds in the experiment with nitrogen gas continued to germinate. Explain why the liquid does **not** move.

....

.... **(2 marks)**

Control of the heartbeat

1 The electrical activity of a heart was studied by stimulating the sinoatrial node (SAN). The time taken for electrical activity to be detected in the atrioventricular node (AVN) was recorded. The time taken for electrical activity to be detected in the SAN at the start of the next cycle was also recorded.

Area of the heart	Time when electrical activity was detected/seconds
AVN	0.15
SAN	0.75

(a) Describe what happens in the atria during the 0.15 seconds before the AVN shows any electrical activity.

> Remember the SAN produces electrical activity intrinsically.

....

....

....

.... **(3 marks)**

Maths skills

(b) Use the information in the table to calculate how often the SAN would show electrical activity in 1 minute. Show your working.

> The avn 'fires' every 0.75 of a second. If it was once per second the number in a minute would be 60. This is easy but think how you actually quickly get this answer the apply the same reasoning to 0.75 sec.

Answer: .... **(2 marks)**

2 The diagram shows structures in the heart that are concerned with the coordination of contraction. Name structures A, B and C.

A is ....

B is ....

C is .... **(3 marks)**

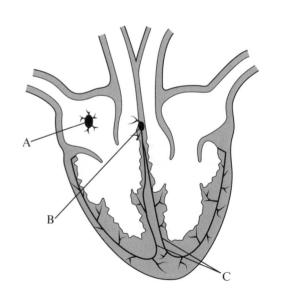

Cardiac output and ventilation rate

1 Describe the changes in the heart that bring about an increase in cardiac output.

> First remind yourself what cardiac output, heart rate and stroke volume are.

....

....

....

....

....

....   **(4 marks)**

Practical skills

2 An experiment was carried out to investigate the effect of a 10-day training programme on a group of 10 individuals. The heart rate and stroke volume of each individual was measured during exercise, both before and after the training programme. The mean results are shown in the table.

Measurement	Before training programme	After training programme
mean heart rate/ beats min^{-1}	152	142
mean stroke volume/cm^3	85	96

(a) State **two** factors that need to be considered when choosing the 10 individuals.

....

....   **(2 marks)**

Maths skills

(b) Using the data in the table, calculate the mean increase in the cardiac output after completing the training programme. Show your working.

> Note the units in which you have to express your answer.

Mean increase = ... dm^3 min^{-1} **(3 marks)**

Maths skills

(c) The data show the volume of blood in a ventricle at different times. Use the data to calculate the cardiac output.

Time	Volume of blood/cm^3
0.0	120
0.1	148
0.2	100
0.3	75
0.4	56
0.5	55
0.6	100
0.7	123

Cardiac output = ...   **(2 marks)**

Spirometry

Practical skills

Maths skills

1 The spirometer trace shown was recorded when an adult was at rest. Calculate the resting breathing rate and tidal volume of the adult.

> Always make sure to include units in your answers to calculations.

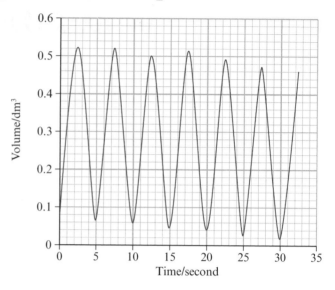

Resting breathing rate = ...

Tidal volume = ...   **(2 marks)**

Maths skills

2 The graphs show spirometer traces for a person at rest and during exercise.

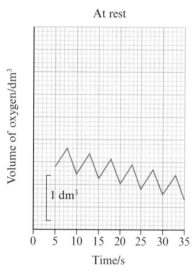

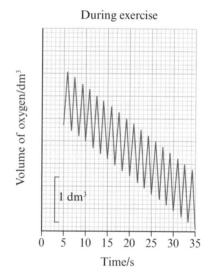

(a) Calculate the increase in minute volume when the person exercises.

Answer: ....   **(3 marks)**

(b) Calculate the oxygen consumption of the person during exercise.

> Oxygen consumption is given by the slope of the curve. At 6 seconds the peak is just over 3.5 dm³.

Answer: ....   **(3 marks)**

Fast and slow twitch muscle

1   An investigation was carried out into the effect of pH on the contraction of muscle fibres. Single muscle fibres were used with their surrounding membranes removed. These fibres will contract when exposed to calcium ions in solution. Isolated slow twitch and fast twitch fibres were tested at pH 7 and pH 6, in a range of calcium ion concentrations. Results for both types of fibre are shown in the graphs.

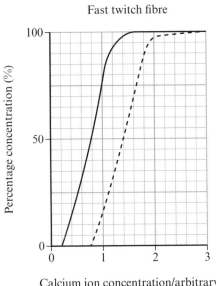

Fast twitch fibre

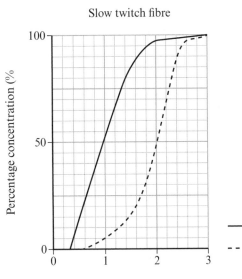

Slow twitch fibre

——— ph 7

- - - - ph 6

(a)   The sensitivity of a muscle fibre is defined as the concentration of calcium ions causing 50% of full contraction. Analyse the data in the graphs to compare and contrast the effect of pH on slow twitch and fast twitch fibres.

> You should conclude that some features of the response are the same for both but other features are different.

....

....

....

....

....

....
                                                     (4 marks)

(b)   Explain how the different responses of these two types of fibre to pH may be related to their different functions in muscle.

....

....

....
                                                     (3 marks)

2   The skeletal muscles of elite sprinters are likely to have many fast twitch muscle fibres. Explain why these muscles are less red in colour than muscles with many slow twitch muscle fibres.

....

....
                                                     (2 marks)

Homeostasis

1 Cortisol is a hormone secreted by the adrenal gland and has many functions in the body. The diagram shows how the secretion of cortisol is controlled. Analyse the information in the diagram and explain how the control of cortisol secretion illustrates the principle of negative feedback.

> The word 'analyse' means that you need to comment on the information and then relate your comments to the situation being discussed to make a judgement.

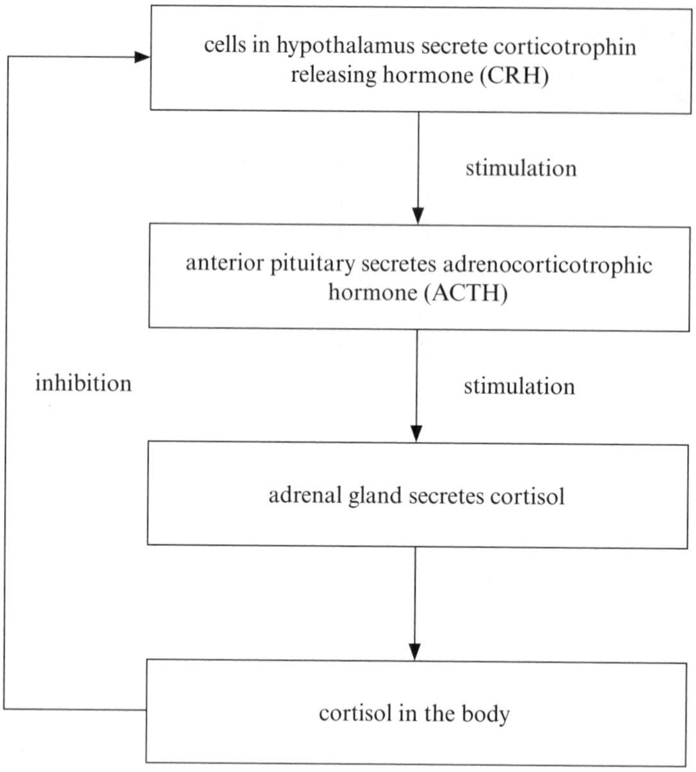

```
            ┌─────────────────────────────────────────────────┐
            │   cells in hypothalamus secrete corticotrophin   │
            │           releasing hormone (CRH)                │
            └─────────────────────────────────────────────────┘
                                    │
                              stimulation
                                    ↓
            ┌─────────────────────────────────────────────────┐
            │   anterior pituitary secretes adrenocorticotrophic│
            │                hormone (ACTH)                     │
            └─────────────────────────────────────────────────┘
                                    │
   inhibition                  stimulation
                                    ↓
            ┌─────────────────────────────────────────────────┐
            │          adrenal gland secretes cortisol         │
            └─────────────────────────────────────────────────┘
                                    │
                                    ↓
            ┌─────────────────────────────────────────────────┐
            │              cortisol in the body                 │
            └─────────────────────────────────────────────────┘
```

....

....

....

.... **(3 marks)**

> Guided

2 Explain how carbon dioxide is involved in the control of breathing rate during exercise.

> Look at page 111 of the Revision Guide.

Carbon dioxide increase or decrease in plasma is detected

by chemoreceptors ....

....

....

....

.... **(4 marks)**

Thermoregulation

Guided

1 Describe the mechanisms that prevent the core body temperature falling below normal.

The fall in temperature is detected by receptors in the hypothalamus.

....

....

....

....

.... **(4 marks)**

2 The structure involved in thermoregulation may cause sweat glands to release more sweat. Explain how increased sweating is involved in the regulation of body temperature.

It is not warm sweat that leaves the body but water vapour which has been created using energy from the blood.

....

....

.... **(3 marks)**

3 Explain how negative feedback controls human body temperature.

Here you must apply a generic understanding of negative feedback to a particular example.

....

....

....

....

.... **(5 marks)**

4 This image of a human head and neck shows part of the CNS. Give the letter and name of the structure involved in thermoregulation.

Although a relatively simple question, it is synoptic as it draws on knowledge of brain structure as well as thermoregulation.

.... **(2 marks)**

Exercise

1 *Doing too little exercise can lead to health problems but too much exercise can also be harmful. Discuss the benefits and potential dangers of exercise in humans.

> In starred questions (*), structure your answer logically showing how the points you make are related to or follow on from each other. You need to select and apply relevant knowledge of biological facts or concepts to support the argument.

....

....

....

....

....

....

....

....

....   **(6 marks)**

2 Explain why obesity is a problem for society.

....

....

....

....   **(3 marks)**

3 Explain why a very efficient metabolism leads to an increased chance of obesity.

....

....

....   **(2 marks)**

Sports participation and doping

1 Sports injuries can result in damaged knee joints. The damaged joint can be repaired using keyhole surgery. The diagram shows a knee joint.

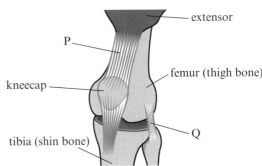

(a) Identify each of the following structures.

   (i) Structure P   (ii) Structure Q

☐ **A** Cartilage   ☐ **A** Cartilage

☐ **B** Ligament   ☐ **B** Ligament

☐ **C** Muscle   ☐ **C** Muscle

☐ **D** Tendon   ☐ **D** Tendon   **(2 marks)**

(b) Joint injuries often shorten the career of athletes. Explain the benefits of keyhole surgery to repair joint damage.

....

....

....

.... **(3 marks)**

2 Research has been carried out on the effect of amphetamines on sporting performance. The table below shows data from a study of volunteers.

Event	Average Performance		Percentage (%) improvement
	After taking placebo/seconds	After taking amphetamines/seconds	
Freestyle	136.88	135.94	0.69
Backstroke	159.80	158.32	0.93
Breastroke	171.87	170.22	–

> Always with percentage calculations do a rough guesstimate first.

(a) Calculate the percentage improvement for breaststroke.

.... **(1 mark)**

(b) Discuss whether the use of amphetamines as a performance-enhancing drug should be made legal.

> In 'discuss' questions, you are likely to have to put both sides of an argument as here.

....

....

....

.... **(3 marks)**

Exam skills

1   *Respiration is a metabolic process which consists of many steps in a metabolic pathway.

> This question is **not** about details of respiration (which you should know) but about the role of enzymes in a metabolic pathway which is a context. In this case, the pathway is respiration but it could just as well be an unnamed pathway and the intermediates given letters.

(a)   The diagram shows three steps in respiration. Each box represents a different substance and each step involves an enzyme. Explain the functions of enzymes in this metabolic process.

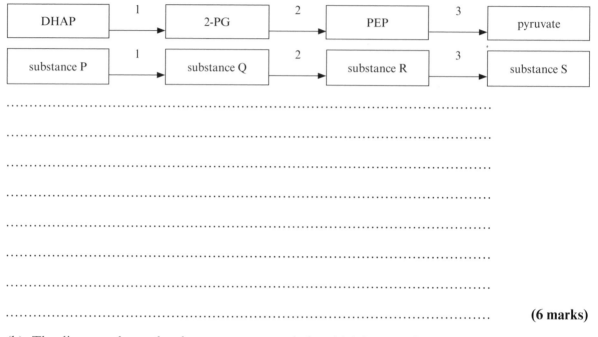

....

....

....

....

....

....

....

....   **(6 marks)**

(b)   The diagram shows the electron transport chain which is part of aerobic respiration. Name substance W and explain how it is formed.

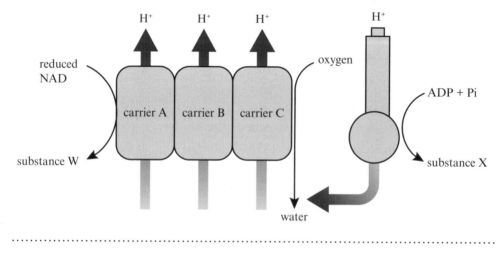

....

....

....   **(3 marks)**

(c)   Name substance X. Explain the link between the formation of substance X and the H$^+$ shown on the diagram.

....

....

....   **(3 marks)**

Mammalian nervous system

1   The table shows features of three types of neurone in a spinal reflex. Place a tick in the box if the feature is present or a cross in the box if absent.

> Always follow instructions very carefully as to how to answer the question; in this case, ticks **and** crosses are required.

Feature	Type of neurone		
	Sensory	Relay	Motor
myelinated			
central cell body			
terminal cell body			
partially surrounded by Schwann cells			
found only in CNS			

(3 marks)

2   The diagram shows part of a nerve cell pathway involved in the reflex.

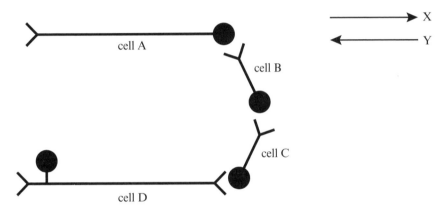

(a) Which of X or Y shows the correct direction of travel of a nerve impulse in cell A?

Answer: ....

(1 mark)

(b) State the different types of neurone shown by cells A to D.

....

....

....   **(3 marks)**

3   Draw a diagram to show the subdivisions of the peripheral nervous system.

> When asked to draw a diagram, make sure you include clear and accurate labels and use a ruler where necessary.

(3 marks)

Stimulus and response

1 The photograph shows the human eye with the black pupil in the centre. The pupil can change size to allow either more or less light into the eye. Its size is controlled by the iris muscles surrounding it.

(a) Explain why the pupil appears black.

.... **(1 mark)**

(b) The pupil increases in diameter in dim light. Explain how neurones enable this response to occur.

> Notice you are being asked to describe what happens in dim light so you need to be very careful about which iris muscle will contract and which one will relax.

....

....

.... **(3 marks)**

2 The graph shows the effect of light intensity on the area of the pupil in the human eye.

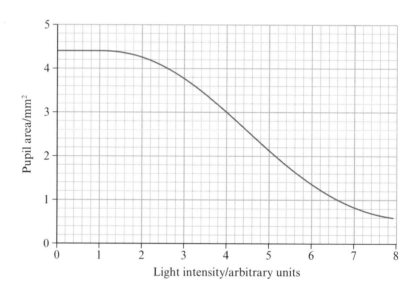

(a) Analyse this information to describe the effect of changing light intensity on the area of the pupil.

> Notice that here you are simply being asked to 'describe' so there is no need to attempt any explanation of what is going on.

....

....

(2 marks)

(b) Change in area of the pupil is controlled by a reflex action. In this pupil reflex, name the receptor and the effector.

....

.... **(2 marks)**

The resting potential

1 The potential difference across the membrane of a neurone was investigated before and after stimulation.

Time/ms	Potential difference/mV
0.00	−70
1.00	−70
1.25	0
1.50	+30
1.75	0
2.00	−80

(a) State the resting potential of this neurone.

Answer: ....   **(1 mark)**

(b) Describe how this resting potential would have been measured.

....

....

....   **(2 marks)**

(c) Explain how this resting potential is maintained.

....

....

....

....

....

....   **(4 marks)**

> Remember that just because a membrane is permeable it does **not** mean there will be movement. This will only happen if there is either a concentration or an electrochemical gradient.

(d) Explain how the potential difference across the membrane is returned to the resting level after depolarisation.

....

....

....

....   **(3 marks)**

Action potential

1 The graph shows a recording of an action potential.

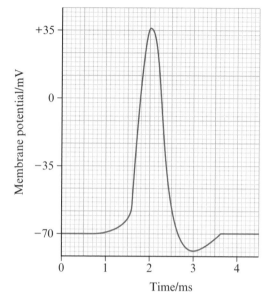

(a) Use the graph to state the time at which the sodium channels open to allow an increased flow of sodium ions into the neurone.

> Here you are being asked the time at which the channels open so a precise time is required, not a range. If the range was required it would say 'are open'.

Answer: .... **(1 mark)**

(b) Use the graph to state the time at which the hyperpolarisation is at its greatest.

Answer: .... **(1 mark)**

Maths skills

(c) Calculate the number of action potentials that could occur in one second if the stimulus is maintained. Show your working.

Answer: .... **(4 marks)**

2 The diagram shows changes in potential difference across the membrane of a neurone during an action potential.

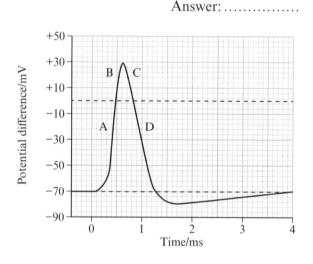

(a) State which ions are able to move across the membrane at positions A and D on the diagram.

....

.... **(2 marks)**

(b) Give an explanation for the movement of ions at position C on the diagram.

....

....

.... **(3 marks)**

Propagation of an action potential

1 Explain how myelination of neurones is an advantage in a reflex pathway.

....

....

....

....

....   **(4 marks)**

2 Explain how impulses are conducted along a myelinated axon after the initiation
of an action potential.

....

....

....

....

....

....   **(5 marks)**

Maths skills

3 The graphs show the conduction velocity and diameter of myelinated and
unmyelinated axons. Compare the percentage increase in velocity in each type
as the diameter doubles.

> You need to choose 2 points on each graph, one double the diameter of the other,
> and calculate the percentage increases and then comment on your results.

....

....

....

....   **(3 marks)**

Synapses

1   Bungarotoxin can be isolated from the venom of a snake called the Prugasti krait.
   In small amounts it can cause paralysis of the diaphragm and intercostal muscles
   by its effects at synapses. Explain how bungarotoxin might cause these effects.

   ....

   ....

   ....

   ....

   ....

   ....   **(4 marks)**

2   *Explain the sequence of events that occurs at the synapse
   after a neurotransmitter has been released.

   > Aim to give a detailed explanation,
   > giving evidence to support your
   > points. Make sure your answer is
   > clear and logically structured and
   > that your points are linked.

   ....

   ....

   ....

   ....

   ....

   ....

   ....

   ....

   ....

   ....

   ....

   ....   **(9 marks)**

3   Atropine is used to dilate the pupil to allow the eye to be examined more easily.
   Atropine inhibits the activity of acetylcholine. Explain how atropine might cause
   this inhibition.

   ....

   ....

   ....   **(2 marks)**

Vision

1   The response to light in humans involves rod cells as receptors.

(a) Place a cross in the box below the diagram that shows the direction light
takes when it stimulates a rod cell.

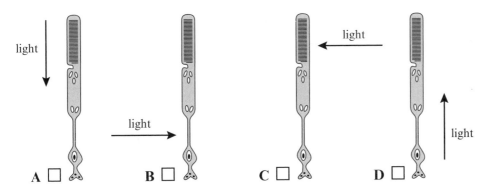

(1 mark)

(b) What happens when a molecule of rhodopsin is bleached by light?

☐ A   opsin changes to retinal

☐ B   retinal changes to opsin

☐ C   trans-retinal changes to cis-retinal

☐ D   cis-retinal changes to trans-retinal   **(1 mark)**

(c) Bleaching of rhodopsin leads to hyperpolarisation of the rod cell membrane.
Which of the following is a description of what happens during hyperpolarisation?

☐ A   Sodium ion channels close while the sodium ion pump stops working.

☐ B   Sodium ion channels close while the sodium ion pump continues to work.

☐ C   Sodium ion channels open while the sodium ion pump continues to work.

☐ D   Sodium ion channels open while the sodium ion pump stops working.   **(1 mark)**

 2   Write an equation to show the changes in rhodopsin in the presence of light.

(2 marks)

 3   Explain how rhodopsin is involved in the conversion of light energy into
electrical energy.

Rhodopsin consists of retinal and opsin ....

....

....

....

....   **(4 marks)**

Plant responses

1 Phytochromes are pigments found in plants. One form of phytochrome is known as P_{FR} (or P_{730}).

(a) Name **one** place in a plant where P_{FR} is found.

.... **(1 mark)**

(b) State the effect that darkness and far-red light have on P_{FR}.

....

.... **(2 marks)**

(c) Describe how the effects of exposure of P_{FR} to darkness could be reversed.

.... **(1 mark)**

2 *A student investigated the effect of natural IAA and artificial IAA on shoot growth. The diagram shows how she set up her investigation.

> In starred questions (*), structure your answer logically showing how the points you make are related to or follow on from each other. You need to select and apply relevant knowledge of biological facts or concepts to support the argument.

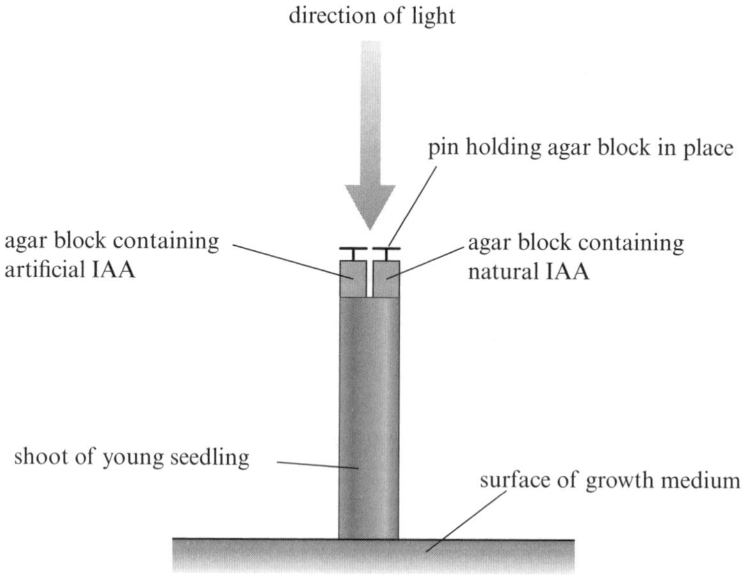

direction of light

pin holding agar block in place

agar block containing artificial IAA

agar block containing natural IAA

shoot of young seedling

surface of growth medium

After 48 hours, the student recorded her observations of the growth of the shoots and concluded that both natural and artificial IAA affected growth and that the artificial IAA had a greater effect than the natural IAA. Analyse this information to explain her conclusions.

> You need to read the student's conclusions carefully and then think about what she must have seen to come to these.

....

....

....

....

....

....

.... **(6 marks)**

Nervous and hormonal control

1 State **four** ways in which hormonal control differs from nervous control.

Make it very clear which one you are talking about in each of the four ways. In some cases it might be necessary to state the situation in both.

....

....

....

.... **(4 marks)**

2 Explain how a hormone affects a target cell.

....

....

....

.... **(3 marks)**

3 Which of the following statements is correct about the sympathetic nervous system?

☐ **A** It is a branch of the somatic system and prepares the body for fight or flight.

☐ **B** It is a branch of the autonomic system and prepares the body for fight or flight.

☐ **C** It is a branch of the somatic system and prepares the body for rest and digest.

☐ **D** It is a branch of the autonomic system and prepares the body for rest and digest. **(1 mark)**

4 Which of the following is true in relation to the length of time over which the hormones act?

☐ **A** insulin > adrenaline > testosterone

☐ **B** adrenaline > testosterone > insulin

☐ **C** testosterone > insulin > adrenaline

☐ **D** All act over the same length of time. **(1 mark)**

The human brain

1 Computed tomography (CT) and functional magnetic resonance imaging (fMRI) are used to investigate brain structure and function. The CT scan shows a human brain with an abnormal area. Arrows indicate this area.

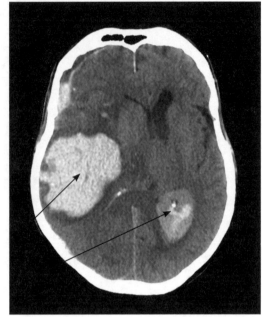

(a) Describe what sort of information this image provides for doctors, so they can decide appropriate treatment of any abnormalities found.

> Look at the image and decide what it tells you about the abnormality and see if you can translate this into what a doctor might get from it.

....

....

....   **(3 marks)**

(b) Describe how fMRI is used to investigate brain function.

> In your discussion of how this technique works, avoid using non-scientific terms such as 'the brain lights up'.

....

....

....   **(3 marks)**

(c) For each of these activities, indicate the region of the brain, W, X, Y or Z, which will be most involved:

regulating core temperature   ....

climbing stairs   ....

regulating carbon dioxide in the blood   ....

choosing a gift   ....   **(4 marks)**

Critical window for development

1 Hubel and Weisel studied the development of vision during the critical window (critical period) of various mammals.

(a) In one investigation, kittens were used.

(i) Explain why kittens were used to study the development of vision in humans.

....

....

....   **(2 marks)**

(ii) Explain why the kittens used were all from one set of parents.

....

....

....   **(2 marks)**

(b) A kitten had its right eye covered for the first seven weeks after birth. The right eye was then uncovered. The left eye was not covered at all. After seven weeks the visual cortex of this kitten was studied.

> Notice that here you are simply being asked to 'describe' so there is no need to attempt any explanation of what is going on.

(i) Describe what happens to the visual pigment in a rod cell when stimulated by light.

....

....

....   **(2 marks)**

(ii) Explain what happened to the visual cortex when the right eye of this kitten was covered for the first seven weeks after birth.

....

....

....

....   **(3 marks)**

(c) Give **one** reason why some people believe that it is ethically unacceptable to use kittens in medical research.

....

....   **(1 mark)**

Learning

1 An investigation was carried out to study habituation in a group of people.

- Each person was played a sound once.
- These sounds made each person blink their eyes and the degree of contraction of one muscle was recorded.
- This was repeated with the sound played 5 times, 10 times, 15 times and 20 times.

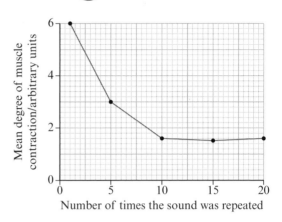

Analyse the data to explain the change in the mean degree of muscle contraction.

The mean degree of muscle contraction decreases ....

....

....

....

(4 marks)

Practical skills

2 A student investigated habituation in snails. The student selected a snail (A) and placed it on a glass plate. The student recorded the time it took for the snail to re-emerge from its shell after tapping with a rod. The plate was tapped every 2 minutes for a total of 12 minutes. The investigation was repeated with two other snails (B and C). The results are shown.

(a) Plot a suitable graph of the mean data.

Tap time/min	Time taken to re-emerge (and start moving)/seconds			
	A	**B**	**C**	**Mean**
0	90	108	80	93
2	40	60	48	49
4	30	40	80	50
6	10	15	20	15
8	0	5	0	2
10	0	0	0	0
12	2	0	0	1

(4 marks)

(b) A Spearman rank correlation test was carried out. The value was −0.93. Using the table, analyse the data to explain the results.

> Degrees of freedom and critical value would not appear in the actual question; they are here to help you.

degrees of freedom

Number of pairs of values (n)	4	5	6	7	8	9	10	11	12
Critical values		1	0.9	0.83	0.79	0.74	0.68	0.65	0.61	0.59

critical value

....

....

....

....

(4 marks)

Brain development

1   Both nature and nurture contribute to brain development. Studies of people's perception of the Müller-Lyer illusion were carried out to test the hypothesis that culture may influence visual perception. The study was carried out on 60 black and 60 white children from a town in the USA and 72 black children from a large town in Zambia, together with 65 from a rural part of Zambia. Each child was shown the illusion 13 times and asked to state if the lines are the same length or a different length.

Müller-Lyer illusion

The results of the study are shown in the graph.

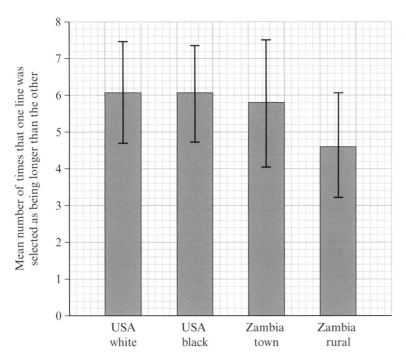

(a) Explain why both black and white Americans were studied.

....

....

.... **(2 marks)**

(b) Deduce whether these results support the hypothesis that incorrect perception of the Müller-Lyer illusion is positively correlated with degree of skin pigmentation.

'Deduce' means you need to work something out.

....

....

.... **(2 marks)**

Brain chemicals

1 L-Dopa can be used to treat people with Parkinson's disease. Using L-Dopa for a long period of time can have side effects that include uncontrolled movement of limbs. It is possible that increasing the levels of serotonin in the brain could be an effective treatment for these side effects. It has been suggested that MDMA (ecstasy) could be used to increase levels of serotonin.

(a) Explain why L-Dopa is used to treat people with Parkinson's disease.

....

....

.... **(2 marks)**

(b) Explain how MDMA could affect levels of serotonin in the brain.

....

....

....

.... **(3 marks)**

(c) In trials of this treatment, marmosets (small monkeys) were given a drug to reduce dopamine production. They were then treated with L-Dopa until they showed the side effects observed in the treatment of people with Parkinson's disease.

(i) Give a reason why the marmosets were treated with a drug to reduce dopamine production.

.... **(1 mark)**

(ii) Describe the ethical issues involved in the use of animals in a trial of this kind.

> In questions about ethical issues always try to give a balanced answer and not just your own opinion.

....

....

.... **(2 marks)**

Guided

(d) The results of the study showed that MDMA did reduce the side effects in the marmosets. Describe the steps that would need to be taken before a similar treatment could be used in humans.

First of all, a small sample should be tested to check for safety.

....

....

....

.... **(4 marks)**

HGP – The Human Genome Project

1 Begun formally in 1990, the Human Genome Project was a 13-year effort designed to identify all of the approximately 25 000 genes in human DNA and determine the sequence of the approximately 3 billion base pairs that make up human DNA. As this knowledge grows, not only does the scope of what might be achieved increase but also the concerns about where it will all lead.

State what is meant by the term genome in the above passage.

....

....

.... **(2 marks)**

2 (a) Give reasons why it would be an advantage for a woman to know that she has a high risk of developing breast cancer.

....

....

....

.... **(3 marks)**

(b) Give reasons why a patient might prefer **not** to know that she has genes that increase her risk of developing breast cancer.

> Remember from your work at AS level that human beings are notoriously bad at understanding the consequences of risk.

....

....

....

.... **(3 marks)**

3 Explain why knowledge of the genome of a person may help to personalise their medical treatment.

....

....

....

....

....

....

.... **(5 marks)**

Genetically modified organisms

1 Cotton plants are used to produce the fibre from which cotton cloth is made. They are grown in certain parts of the world, such as India and the USA, where cotton farming is an important way in which people earn their living. Normally, cotton plants need to be sprayed with chemical insecticides to kill insect pests. Recently, genetically modified (GM) cotton plants have been developed which produce a natural insecticide of their own. This insecticide kills the insect pests but is harmless to humans.

(a) Discuss advantages to cotton farmers and to the environment of growing genetically modified cotton.

....

....

....

.... **(3 marks)**

(b) Compare and contrast the use of genetic modification of plants and conventional plant breeding to improve crop yield.

> Make sure you write about ways in which the two processes are similar and ways in which they are different.

....

....

....

....

.... **(4 marks)**

(c) *In some countries genetically modified organisms are banned; in others they are freely grown. Evaluate the science and ethics of banning the use of GMOs.

> Make sure you structure your answer logically, and show how your points follow on from each other. Aim to support your points with relevant biological facts and/or evidence. In ethical questions like this, you should always look at both sides of the argument.

....

....

....

....

....

....

....

....

....

....

.... **(9 marks)**

Exam skills

1 The diagram shows a synapse. With reference to the numbers in the diagram, explain how a change in the functioning of the synapse might bring about habituation to a stimulus.

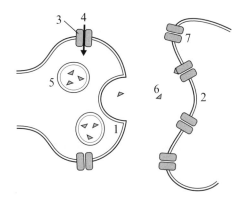

> This is a question that asks you to describe a sequence of events started off by one event. You must link your answer to the diagram throughout.

....

....

....

....

....

....

....   **(5 marks)**

2 The diagrams below show the effects of internal calcium ion (Ca^{2+}) concentration on the contraction of an isolated muscle fibre.

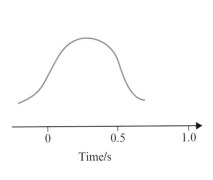

Time/s

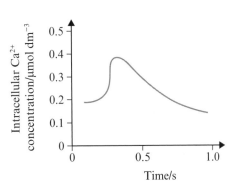

Time/s

(a) Explain how the release of calcium ions in the muscle fibre is stimulated.

....

....

....

....   **(2 marks)**

(b) Explain the role of calcium ions in muscle contraction.

....

....

....

....   **(3 marks)**

Edexcel publishes official Sample Assessment Material on its website. This timed test has been written to help you practise what you have learned and may not be representative of a real exam paper.

Remember that the official Edexcel specification and associated assessment guidance materials are the only authoritative source of information and should always be referred to for definitive guidance.

AS Level Timed Test Paper 1: Lifestyle, Transport, Genes and Health

Time: 2 hours

Where you see *, you will be assessed on your ability to provide a well-structured answer with clearly linked points.

---

1 The cell surface membrane is involved in the transport of materials into and out of the cell.

The symbols below represent some of components of a cell surface membrane.

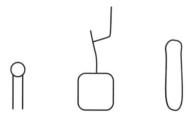

(a) Draw a diagram to show the complete structure of part of a cell surface membrane using these symbols. **(3)**

(b) The graph below shows the rate of uptake of a substance by facilitated diffusion into a cell.

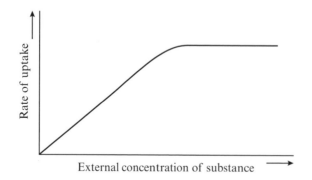

Explain why the rate changes as shown. **(2)**

(Total for question 1 = 5 marks)

2 Blood vessels are adapted to their function.

    (a) (i)   Draw and label a diagram to show the structure of an artery. **(3)**

        (ii) Explain how the structure of an artery relates to its function. **(2)**

    (b) Explain why a blood clot in an artery leading to the brain could cause a stroke. **(3)**

    (c) A number of factors have been identified that increase the risk of CVD.

        One of these factors is genetic. The genotype of some individuals causes them to be more at risk of developing CVD. One gene that influences this risk is the KIF6 gene. Carriers of the 719 Arg allele of this gene are more at risk of CVD.

        (i)   Explain the meaning of the term genotype in relation to 719 Arg. **(1)**

        (ii) Explain the meaning of the term allele in relation to 719 Arg. **(1)**

    (d) Trials have shown that plant statin therapy is more effective in 719 Arg carriers than in non-carriers of this allele. **(2)**

(Total for question 2 = 12 marks)

3 Invertase is an enzyme use to create soft centres in chocolates. It does this by breaking down crystalline sucrose encased in chocolate over a period of a few days.

    (i)   State how the conditions could be adjusted to make sure the breakdown takes a few days. **(1)**

    (ii) State the name of the type of reaction catalysed by invertase. **(1)**

    (iii) State the names of all the reactants and products of this breakdown. **(2)**

(Total for question 3 = 4 marks)

4 The diagram below shows the sequence of events leading to polypeptide synthesis.

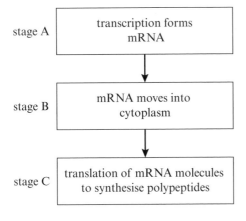

    (a) Place a cross in the box next to the correct term that completes each of the following statements.

        (i)   Translation takes place in or on the **(1)**

            ☐ **A** Golgi apparatus

            ☐ **B** lysosome

            ☐ **C** nucleus

            ☐ **D** ribosome

(ii) A triplet of bases that could not be found in mRNA is (1)

    ☐ **A** adenine; adenine; guanine

    ☐ **B** adenine; thymine; guanine

    ☐ **C** adenine; cytosine; guanine

    ☐ **D** adenine; uracil; guanine

(iii) mRNA is made in the (1)

    ☐ **A** nucleus

    ☐ **B** ribosomes

    ☐ **C** endoplasmic reticulum

    ☐ **D** Golgi apparatus

(b) Explain how proteins are made in cells. (5)

(Total for question 4 = 8 marks)

5 A gene contains 450 bases of which 32% are guanine.

(a) Calculate the number of uracil bases present in the mRNA made from this gene. (3)

(b) DNA is made up of the following components:

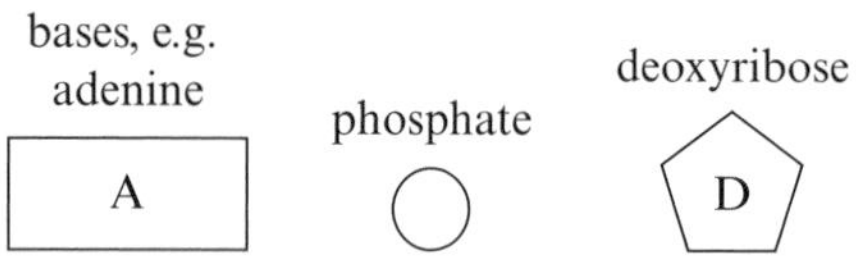

bases, e.g.
adenine   deoxyribose

           phosphate

   ☐ A   ◯   ⬠ D

Use these diagrams to draw the part of the antisense strand of a DNA molecule that would give rise to the codon UAC in mRNA. (3)

(c) In their 1953 paper describing their model of DNA structure, Watson and Crick said: 'It has not escaped our notice that the specific pairing we have postulated immediately suggests a possible copying mechanism for the genetic material.'

Explain the copying mechanism that is now understood. (4)

(Total for question 5 = 10 marks)

6 A triglyceride is one type of lipid.

(a) For each of the descriptions below, put a cross in the box that corresponds to the correct statement about lipids or triglycerides.

(i) Triglycerides are composed of (1)

    ☐ **A** three glycerol molecules and three fatty acid molecules

    ☐ **B** one glycerol molecule and three fatty acid molecules

    ☐ **C** one glycerol molecule and one fatty acid molecule

    ☐ **D** three glycerol molecules and one fatty acid molecule

(ii) The bond between a glycerol molecule and a fatty acid molecule is **(1)**

    ☐ **A** a glycosidic bond

    ☐ **B** a peptide bond

    ☐ **C** a phosphodiester bond

    ☐ **D** an ester bond

(iii) This bond is formed by **(1)**

    ☐ **A** hydrolysis

    ☐ **B** condensation

    ☐ **C** a chain reaction

    ☐ **D** an automatic reaction

(iv) Unsaturated lipids **(1)**

    ☐ **A** do not have any double bonds

    ☐ **B** have double bonds only between carbon atoms

    ☐ **C** have double bonds between carbon atoms and between carbon and oxygen atoms

    ☐ **D** have double bonds only between carbon and oxygen atoms

(b) Adult volunteers took part in an investigation to find out the effect of dietary changes on their risk of developing coronary heart disease.

In this investigation, 5% of the volunteers' energy intake was changed from one food source to another. The volunteers' total energy intake remained constant.

The graph below shows the results of this investigation.

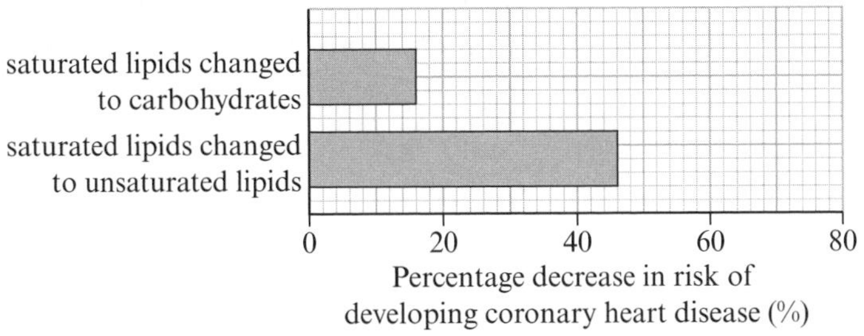

(i) Explain why it was necessary to ensure that their total energy intake remained constant. **(2)**

(ii) Explain the results of this investigation. **(3)**

(Total for question 6 = 9 marks)

7   Fenitrothion is an insecticide used to control insects that feed on crop plants.

Before an insecticide is approved for use, its effects on insects and other animals are tested. The testing of fenitrothion showed that it affects the permeability of animal cell membranes.

Some scientists investigated the effect of fenitrothion on the permeability of the plant cell membranes of beetroot.

The diagram below shows a beetroot cell with a vacuole containing a red pigment called betalain.

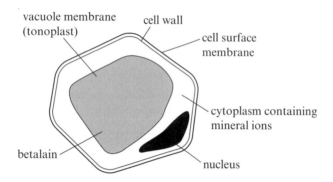

Discs were cut from a beetroot. Any betalain on the outside of the discs was removed by washing the discs in water.

Twenty discs were placed into a beaker containing 20 cm³ of fenitrothion solution in water.

Betalain began to leak from the discs, changing the colour of the solution. The colour of the solution in the beaker was recorded every hour. In another experiment, they investigated the movement of minerals out of the cells over the same time period.

The graph shows their results.

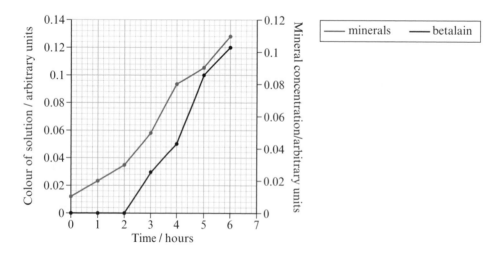

(a) Analyse all the information to compare and contrast the patterns of movement of minerals and betalain pigment.   **(4)**

*(b) A scientist thought pH might affect permeability.

Explain how this experiment could be modified to determine the pH value that would have the least or no effect on permeability of the membranes.   **(6)**

(Total for question 7 = 10 marks)

8 Obesity is a significant problem in western countries and an increasing problem in some other parts of the world.

The graph below shows the percentage of the male population (M) and the female population (F) who are either overweight or obese in five different countries.

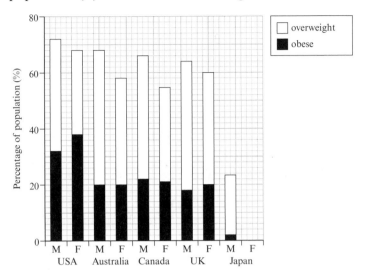

(a) The graph shows that in Japan 2% of the male population are obese and 23% are overweight. In the female population of Japan, 2% are obese and 16% are overweight. Plot this information on the graph above. **(3)**

(b) For the statement below, put a cross in the box that corresponds to the correct statement.

    (i) The graph shows that **(1)**
       ☐ **A** a higher percentage of males than females are overweight
       ☐ **B** a higher percentage of females than males are overweight
       ☐ **C** there is no correlation between being overweight and gender
       ☐ **D** an equal percentage of males and females are overweight

    (ii) State the country with the highest percentage of males who are obese. **(1)**

    (iii) State the country with the same percentage of females as the UK who are overweight. **(1)**

    (iv) Calculate the ratio of overweight males to females in the USA. **(1)**

(c) Explain why it would be incorrect to conclude that, in Japan, the same number of males as females is obese. **(2)**

(d) The graph below shows the consumption of three types of sweetener in the USA from 1966 to 2002.

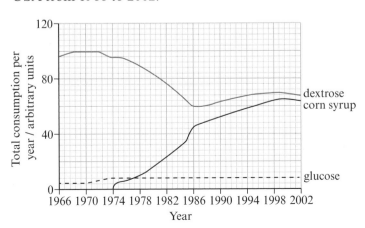

From 1976, the number of obese people in the USA started to increase rapidly. It was suggested that there was a correlation between the consumption of corn syrup and obesity.

(i) Explain the meaning of the term correlation. **(1)**

(ii) Analyse the information in part (d) to explain the evidence that suggests there is a correlation between the consumption of corn syrup and obesity. **(3)**

(Total for question 8 = 13 marks)

9 Lipase can be used in the manufacture of biodiesel. The graph shows the course of a reaction at one enzyme concentration as part of an experiment to look at effect of enzyme concentration on rate for this reaction. The table shows initial rates for all lipase concentrations in the experiment.

(a) Deduce the concentration in the table that produced the results in this graph. Show your working. **(3)**

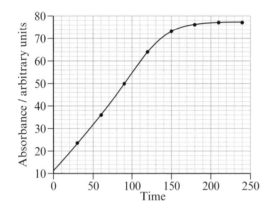

Enzyme concentration (%)	Initial rate of reaction / arbitrary units s^{-1}
0.0	0.00
0.5	0.20
1.5	0.44
3.0	0.75
4.5	0.76
5.5	0.79

(b) Plot the data in the table on a graph. **(3)**

(c) Analyse the graph you have plotted to explain the effect of enzyme concentration on reaction rate. **(3)**

(Total for question 9 = 9 marks)

TOTAL FOR PAPER = 80 MARKS

AS Level Timed Test Paper 2: Development, Plants and the Environment

> **Time: 2 hours**
> Where you see *, you will be assessed on your ability to provide a well-structured answer with clearly linked points.

---

1 Carl Woese suggested that living organisms could be grouped into three domains, Bacteria, Archaea and Eukarya.

    (a) (i) Explain the principles underlying Woese's three domain system of taxonomy. **(3)**

        (ii) When Carl Woese first suggested that all organisms could be classified into one of the three domains, his ideas were not accepted.

            Explain how Woese's idea could have been critically evaluated. **(3)**

    (b) Place a cross in the box next to the best definition of a species. **(1)**

        ☐ **A** Individuals can interbreed to produce fertile offspring.

        ☐ **B** Individuals can interbreed to produce hybrid offspring.

        ☐ **C** Individuals can interbreed to produce sterile offspring.

        ☐ **D** Individuals can interbreed to produce viable offspring.

(Total for question 1 = 7 marks)

2 Fertilisation happens in both mammals and flowering plants.

    (a) (i) Compare and contrast the function of mammalian sperm and eggs. **(3)**

        (ii) Explain how, in mammals, events following the acrosome reaction prevent more than one sperm fertilising an egg. **(5)**

    (b) Human stem cell research involves the use of both totipotent and pluripotent stem cells.

        (i) Pluripotent stem cells are **(1)**

            ☐ **A** specialised cells that can differentiate to give rise to almost any type of cell in the body, including totipotent cells

            ☐ **B** specialised cells that can differentiate to give rise to any type of cell in the body, excluding totipotent cells

            ☐ **C** unspecialised cells that can differentiate to give rise to almost any type of cell in the body, excluding totipotent cells

            ☐ **D** unspecialised cells that can differentiate to give rise to any type of cell in the body, including totipotent cells

        (ii) Describe the differences between a totipotent stem cell and a pluripotent stem cell. **(3)**

(Total for question 2 = 12 marks)

3 The diagram shows a β-glucose molecule, which is the monomer of cellulose.

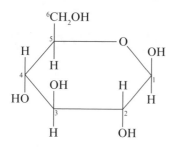

(a) Draw the products that are formed from a condensation reaction between two β-glucose molecules. **(2)**

(b) Explain how the structure of cellulose makes it a suitable molecule to form the walls of cells. **(4)**

(c) A study was carried out in which thin films of cellulose were put in contact with an enzyme, cellulase, which breaks down cellulose. The films were weighed at intervals and the loss in mass, due to cellulose digestion, was recorded. This was done at five pH values: 3, 4, 5, 7 and 10.

The data are shown in the graph.

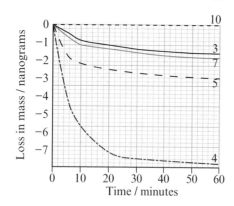

Analyse the data to plot a graph of initial rate against pH. **(4)**

(Total for question 3 = 10 marks)

4 (a) A ten-year survey was carried out on the species richness of hedgerows and roadside verges. The data are shown below.

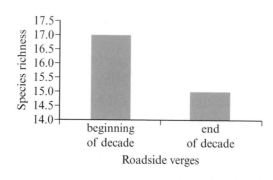
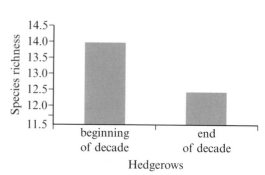

(i) Analyse the data to compare the changes in these two habitats over the ten-year period. **(4)**

(ii) Criticise this study, which was entitled 'Biodiversity changes in hedgerows and roadside verges over a ten-year period'. **(4)**

(b) Seed banks are an effective means of conserving plant species.

Write a short briefing, intended for a government committee, describing how seed banks work and why their funding should be continued. **(3)**

(Total for question 4 = 11 marks)

5 The photograph below shows a waxy leaf frog (*Phyllomedusa sauvagii*). This species of frog is found in hot, dry areas of South America.

It has glands that produce waxy lipids to spread over its skin. This reduces water loss.

The waxy leaf frog is active only at night, when it hunts for insects in the trees.

(a) (i) Describe how the waxy leaf frog is physiologically adapted to its environment. **(1)**

   (ii) Explain how the behaviour of the waxy leaf frog enables it to survive in its habitat. **(2)**

(b) Explain the term niche using the waxy leaf frog as an example. **(2)**

(c) Explain how natural selection could have given rise to the adaptations shown by the waxy leaf frog. **(5)**

(Total for question 5 = 10 marks)

6 Radishes (*Raphanus sativus*) were grown in pots containing washed sand. The pots absorbed water and minerals, via a wick system, from a solution containing minerals.

One solution contained all the minerals a plant needed. Other solutions contained all the minerals with the exception of one in each case.

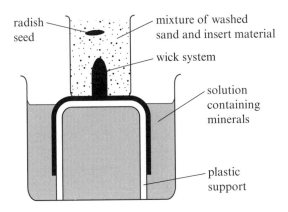

After three weeks, the dry mass of roots and shoots was measured and a mean was calculated. The table shows the results.

Solution	Mean dry mass / mg		
	Shoots	Roots	Total
with all minerals present	235	139	374
without potassium	188	139	327
without iron	231	96	327
without nitrate	141	75	216
without calcium	167	40	207
without magnesium	178	29	207
without phosphate	186	16	202

(a) (i) Plot the data from the table to compare the mean dry masses of shoots and the mean dry masses of roots in each of the solutions. (4)

    (ii) It was concluded that a decrease in root:shoot ratio shows a decline in plant health. Calculate ratios to evaluate this conclusion. (3)

(b) Explain the role of calcium in plants. (2)

(Total for question 6 = 9 marks)

7 (a) In the lac operon system, lactose is the inducer molecule. Which of the following is true of the inducer in this system?

    ☐ **A** It combines with the operator region and activates operons.

    ☐ **B** It combines with repressor proteins and inactivates them.

    ☐ **C** It combines with the beta galactosidase gene and activates it.

    ☐ **D** It directly activates RNA polymerase. (1)

The diagram shows the lac operon as found in *E. coli*.

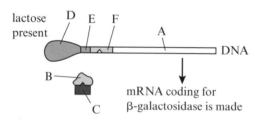

*(b) Explain the sequence of events from the introduction of lactose into the medium in which the bacteria are growing to the production of mRNA using the names of structures A–F where appropriate. (6)

(Total for question 7 = 7 marks)

8 (a) Organisms are made of cells. There are differences in the ultrastructure of prokaryotic and eukaryotic cells. There are also differences between plant and fungal cells.

(i) Ribosomes are found in **(1)**

☐ **A** fungal cells only

☐ **B** eukaryotic cells only

☐ **C** prokaryotic cells only

☐ **D** eukaryotic and prokaryotic cells

(ii) Amyloplasts are found in **(1)**

☐ **A** fungal cells only

☐ **B** plant cells only

☐ **C** prokaryotic cells only

☐ **D** eukaryotic and prokaryotic cells

(iii) Mitochondria are found in **(1)**

☐ **A** fungal cells only

☐ **B** eukaryotic cells only

☐ **C** prokaryotic cells only

☐ **D** eukaryotic and prokaryotic cells

(iv) Centrioles are found in **(1)**

☐ **A** animal cells only

☐ **B** all eukaryotic cells

☐ **C** prokaryotic cells only

☐ **D** eukaryotic and prokaryotic cells

(b) The diagram below shows a growing tip of one fungal hypha containing vesicles, labelled Z. These vesicles contain digestive enzymes.

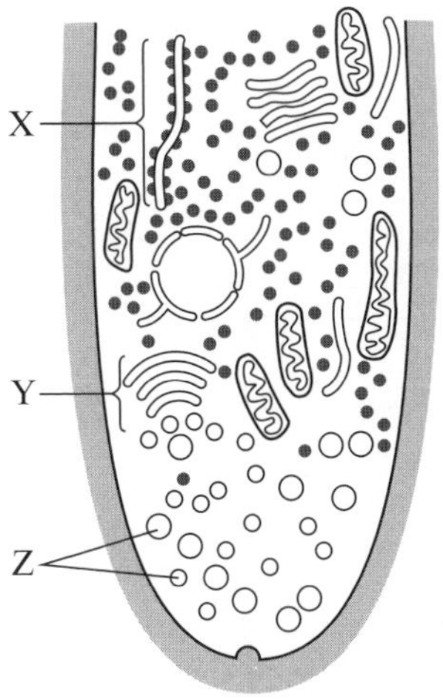

(i) Place a cross in the box next to the correct name of the organelle labelled Y on the diagram. **(1)**

☐ **A** Golgi apparatus

☐ **B** mitochondrion

☐ **C** rough endoplasmic reticulum

☐ **D** smooth endoplasmic reticulum

(ii) The organelles labelled X, Y and Z on the diagram are involved in the synthesis and secretion of digestive enzymes.

Describe the roles of these organelles in the synthesis and secretion of digestive enzymes. **(4)**

(iii) Explain why it is necessary for fungi to produce different enzymes to digest starch and cellulose. **(4)**

(c) The photograph shows part of a cellulose cell wall seen using an electron microscope.

Using the information in the photograph and your own knowledge, describe the structure of a cellulose cell wall. **(2)**

(Total for question 8 = 15 marks)

TOTAL FOR PAPER = 81 MARKS

A Level Timed Test Paper 1: The Natural Environment and Species Survival

> **Time: 2 hours**
>
> Where you see *, you will be assessed on your ability to provide a well-structured answer with clearly linked points.

1 The human body responds to a virus infection by producing interferon and antibodies.

The graph shows the change in the number of virus particles, the level of interferon and the level of antibodies in a person in the weeks following an infection.

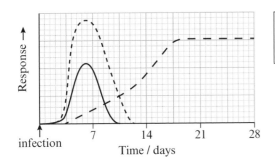

(a) Describe the structure of a virus. **(3)**

(b) (i) Explain why there is a delay, following this infection, before the number of virus particles increases. **(2)**

(ii) Explain the change in the number of virus particles from day 1 to day 5. **(2)**

(c) Explain why there is a delay before the level of antibodies starts to rise. **(4)**

(Total for question 1 = 11 marks)

2 The graphs below show the relationship between net primary production and two abiotic factors.

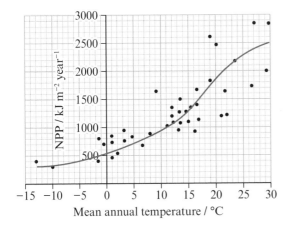 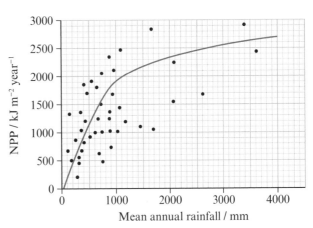

(a) (i) Describe the meaning of the term net primary productivity (NPP). **(2)**

(ii) Analyse the information in the graphs to explain the relationship between NPP and each of these two environmental factors. **(5)**

(iii) Explain the appearance of a graph showing the relationship in ecosystems between gross primary productivity (GPP) and rainfall. **(3)**

(b) The diagram below shows energy transfer between two trophic levels in an ecosystem.

```
┌─────────────────────────┐              ┌──────────────────────────────┐
│      Tropic level 1      │   ─────▶     │        Tropic level 2        │
│ energy in biomass = 5300 kJ │           │ energy of food ingested = 2800 kJ │
│                          │              │   energy lost in respiration,    │
└─────────────────────────┘              │ urine and faeces = 1750 kJ       │
                                         └──────────────────────────────┘
```

Calculate the percentage of energy in trophic level 1 transferred to new biomass in trophic level 2. **(2)**

(Total for question 2 = 12 marks)

3 The diagram shows part of the light-dependent reactions of photosynthesis.

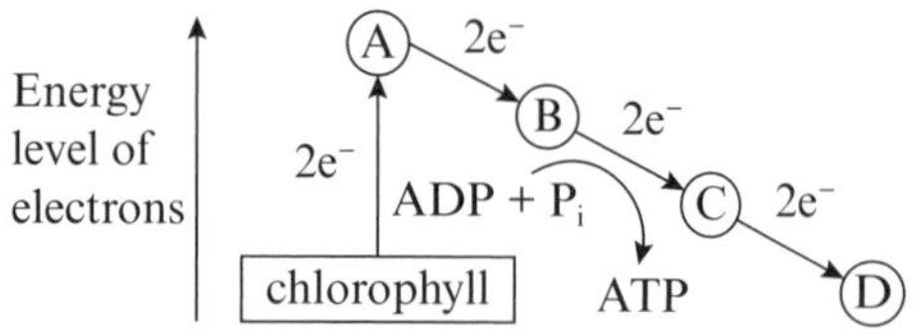

(a) (i) Explain how a pair of electrons moves from chlorophyll to the electron carrier system. **(2)**

(ii) Explain how electrons lost from chlorophyll are replaced. **(2)**

(iii) Describe how ATP is synthesised with reference to the diagram. **(3)**

(b) (i) Describe the Calvin cycle. **(5)**

Put a cross in the box next to the correct answer.

(ii) Products which can be made from glucose with no further elements required include **(1)**
☐ A starch
☐ B amino acids
☐ C nucleotides
☐ D ATP

(iii) Elements which need to be added to glucose to make proteins include **(1)**
☐ A phosphorus
☐ B magnesium
☐ C nitrogen
☐ D calcium

(Total for question 3 = 14 marks)

4 Lipase is an enzyme that breaks down triglycerides into their component molecules.

An investigation was carried out to compare the effect of temperature on the activity of lipase R and lipase S. These lipases were obtained from two different species of bacteria.

The results are shown in the table.

Temperature/°C	Activity of lipase R/ arbitrary units	Activity of lipase S/arbitrary units
30	25	10
40	43	22
50	58	28
60	80	38
70	70	59
80	52	65

(a) (i) State the names of the two products from the complete breakdown of triglycerides by lipase. **(1)**

(ii) Analyse the data graphically to explain the effect of temperature on the activity of lipase R. **(5)**

(b) In this investigation the pH was kept at 7. Further experiments were carried out to find the optimum temperature for the activity of lipase S.

(i) Place a cross in the box next to the correct term that completes the following sentence.
The pH was controlled to make this investigation **(1)**

☐ **A** accurate

☐ **B** precise

☐ **C** repeatable

☐ **D** valid

(ii) Place a cross in the box next to the temperatures that should be used in further experiments with lipase S. **(1)**

☐ **A** 30°C to 80°C

☐ **B** 60°C to 80°C

☐ **C** 70°C to 80°C

☐ **D** 70°C to 100°C

(iii) In the experiments on lipase S temperature is the **(1)**

☐ **A** dependent variable

☐ **B** control variable

☐ **C** independent variable

☐ **D** non-experimental variable

(iv) Put a cross in the box next to the word that best describes the bacterium from which lipase S was extracted. **(1)**

☐ **A** thermophile

☐ **B** acidophile

☐ **C** thermophobe

☐ **D** acidophobe

(Total for question 4 = 10 marks)

5 Following the extraction of coal from the ground in the United Kingdom, the unwanted material was usually deposited in large heaps known as bings. Most of the material in a bing is shale fragments composed of minerals and clay.

There have been a number of studies of the colonisation and the development of plant communities on bings. In these studies, the approximate age of the bing can be estimated by reference to the type of plant community growing on the bing.

This is shown in the table below.

Type of plant community	Approximate age of bing/years
lichens and mosses	3–15
grasses and small herbs	15–40
grasses, small herbs and large herbs	40–70
small trees and shrubs	60–80
large trees, small trees and shrubs	80–more than 100

(a) Place a cross in the box next to the mineral ion that would need to be present if plants, such as grasses and herbs, are to grow successfully on a bing. **(1)**

☐ **A** copper   ☐ **C** sodium

☐ **B** nitrates   ☐ **D** sulfites

(b) Place a cross in the box that describes the gradual change in the type of plant community growing on a bing. **(1)**

☐ **A** endemism   ☐ **C** phylogeny

☐ **B** evolution   ☐ **D** succession

(c) Analyse the information in the table to explain why the type of plant community growing on a bing changes over time. **(6)**

(d) After 100 years, the community on a bing becomes stable.
Explain why this type of community is stable. **(4)**

(Total for question 5 = 12 marks)

6 The sea anemone, *Anthopleura elegantissima*, occupies a niche at the secondary and tertiary consumer levels in a food web on the shores of North America.

At high tide, the sea anemone is active and feeds on a variety of small invertebrate animals and fish. It paralyses its prey using stinging cells on tentacles. The food is then passed into the gut of the sea anemone for digestion by enzymes. The sea anemones also form the food of various carnivores.

At low tide, the sea anemones are exposed on the rocks of the shore where they remain stationary until the water returns at high tide.

During this exposure, the tentacles and body of each sea anemone are contracted into a rounded mass.

(a) Explain what is meant by the term niche, using the sea anemone *Anthopleura elegantissima* as an example. **(3)**

(b) Explain why the anemones contract when exposed at low tide. **(3)**

Line transects were used to study the effects of abiotic factors on the distribution of *A. elegantissima* on a rocky shore. The graph below shows how the temperature, height above sea level at low tide and number m⁻² of *A. elegantissima* varied in a study.

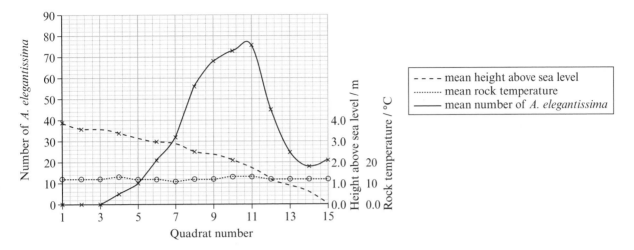

(c) (i) State the type of factors that temperature and height above sea level represent. **(1)**

(ii) Analyse the data to explain the effects of these two factors on the distribution of *A. elegantissima* on this shore. **(3)**

(Total for question 6 = 10 marks)

7 The spruce bark beetle feeds on and breeds in spruce trees. If a large number of beetles are on a spruce tree, the tree will die. These dead trees appear red from the air.

It is thought that the number of beetles is being affected by climate change.

Each year, the extent of damage to forest in Alaska was estimated by measuring the size of the 'red area' from aerial photographs.

The drought index of the woodland was also determined. A high drought index indicates warm, dry conditions and a low drought index indicates cool, moist conditions.

The graphs below show the changes in 'red area', mean summer temperature and drought index in Alaskan forest from 1930 to 2000.

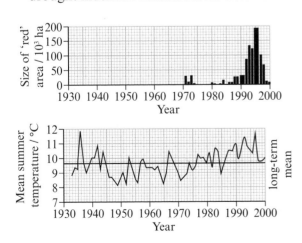

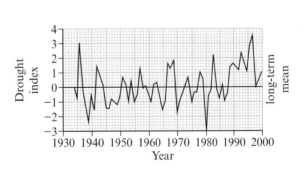

(a) Describe the changes in the size of the 'red area' from 1970 to 2000. **(3)**

(b) Explain why the number of spruce bark beetles is affected by temperature. **(2)**

(c) Analyse the information in the graphs to describe the evidence for climate change being responsible for the size of the 'red area'. **(3)**

(d) Explain why a valid conclusion cannot be made about the effect of climate change on the size of the 'red area'. **(3)**

(Total for question 7 = 11 marks)

8 Experimental plots were set up on a hillside to investigate the possible effects of climate change on biodiversity of limestone grassland.

Three different treatments were used:

- H = heat. Plots with electric heating cables buried underground which maintained a soil temperature 3°C above that of the surrounding hillside during the winter.

- D = drought. Plots with sliding glass panels which automatically moved into place whenever it rained ensuring that the soil received no water throughout July and August.

- W = water. Additional water was added to these plots from June to September keeping the soil moist (i.e. no drought). These plots had neither heating cables nor sliding glass panels.

There were also control plots left untreated but recorded at the same time and in the same way as treatments H, D and W.

The graphs show the percentage change in abundance of six plant species subjected to the three treatments, H = heat, D = drought and W = water.

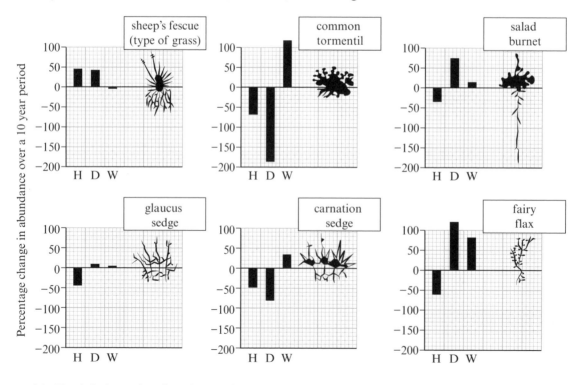

(a) Explain how the abundance of each of the species in each of the three areas may have been determined in order to obtain the data to produce the graphs shown. **(4)**

(b) (i) Assuming that common tormentil had an average abundance of 24% cover at the beginning of the study, calculate the average rate of change of its abundance over the ten years of the study. (3)

(ii) Given that you had the actual data for abundance in all the sample plots from any of the species, explain which statistical test you would use to determine if the difference in abundance between the beginning and the end of the study was significant. (2)

(c) Analyse the data to describe what the results indicate is likely to be the effect on the abundance of salad burnet if the climate were to become both hotter and wetter. (2)

(Total for question 8 = 11 marks)

9 Yeast is a single-celled fungus. It can reproduce asexually by a process called budding.

When the bud is big enough it separates from the original yeast cell.

Yeast can be grown in a culture containing all the nutrients needed for growth. Small samples of the culture can be removed and the yeast observed using a light microscope.

The photograph below shows yeast budding, as seen using a light microscope.

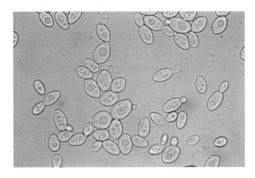

The diagram below shows a yeast cell with a bud.

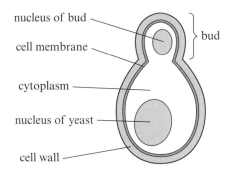

(a) Explain how the properties of the cell membrane enable the yeast cell to form a bud. (4)

(b) Temperature affects the rate of asexual reproduction in yeast.
Devise an investigation that could be carried out to study the effects of temperature on the rate of asexual reproduction in yeast. (5)

(Total for question 9 = 9 marks)

TOTAL FOR PAPER = 100 MARKS

A Level Timed Test Paper 2: Energy, Exercise and Co-ordination

> **Time: 2 hours**
>
> Where you see *, you will be assessed on your ability to provide a well-structured answer with clearly linked points.

---

1 Rod cells in the eye are linked to the brain by neurones.

   (a) Place a cross in the box to identify the answer that correctly completes each statement.

   (i) The pigment in a rod cell is made of opsin and **(1)**

   ☐ **A** retina

   ☐ **B** retinal

   ☐ **C** retina

   ☐ **D** retinol

   (ii) When light stimulates a rod cell the pigment changes. This pigment is **(1)**

   ☐ **A** iodopsin

   ☐ **B** phytochrome far red

   ☐ **C** phytochrome red

   ☐ **D** rhodopsin

   (iii) Once the pigment has changed, the concentration of sodium ions inside the rod cell **(1)**

   ☐ **A** decreases

   ☐ **B** does not change

   ☐ **C** increases

   ☐ **D** reaches equilibrium with the outside of the cell

   (iv) After changing, the pigment takes time to become functional again. This is because **(1)**

   ☐ **A** it has to bleach

   ☐ **B** the membrane has to be polarised

   ☐ **C** the rod cell needs to reset

   ☐ **D** two components have to be rejoined

   (v) The cell that links a rod cell to a sensory neurone is **(1)**

   ☐ **A** a bipolar neurone

   ☐ **B** a multipolar neurone

   ☐ **C** a unipolar neurone

   ☐ **D** an optic nerve

(b) Decreasing the intensity of light entering the eye causes pupil dilation.
Describe the roles of the circular and radial muscles in pupil dilation. **(2)**

(Total for question 1 = 7 marks)

2 Respiration is a metabolic process which consists of many steps.

(a) The diagram below shows a metabolic process consisting of three steps. Each
letter represents a different substance and each number a different enzyme.

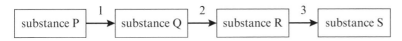

Explain the functions of enzymes in this metabolic process. **(4)**

(b) The diagram shows the electron transport chain, which is part of aerobic
respiration.

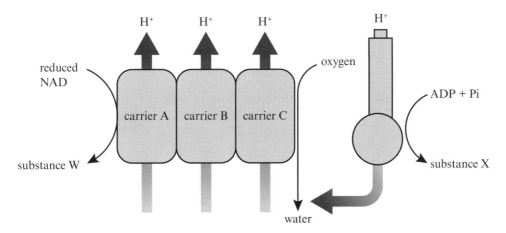

(i)   Describe how substance W is formed. **(3)**

(ii) Explain the link between the formation of substance X and the H⁺
shown on the diagram. **(3)**

(c) The diagram below shows a respirometer used to measure the rate of aerobic
respiration in small organisms.

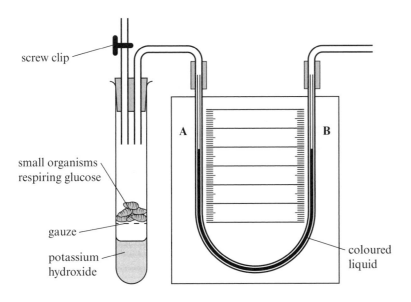

Place a cross in the box that correctly shows the movement of the coloured liquid in the U-shaped tube for each situation shown.

    (i) Screw clip is open   **(1)**

        ☐ **A**   towards A

        ☐ **B**   towards B

        ☐ **C**   does not move

        ☐ **D**   the direction of movement will depend on temperature

    (ii) Screw clip is closed   **(1)**

        ☐ **A**   towards A

        ☐ **B**   towards B

        ☐ **C**   does not move

        ☐ **D**   the direction of movement will depend on temperature.

    (iii) Potassium hydroxide is replaced with water and screw clip is closed   **(1)**

        ☐ **A**   towards A

        ☐ **B**   towards B

        ☐ **C**   does not move

        ☐ **D**   the direction of movement will depend on temperature

The apparatus was used to measure the rate of respiration of germinating seeds in air.
The distance moved by the coloured liquid was measured at 15-minute intervals for one hour.

This was repeated with the air replaced by nitrogen gas.

The rate of respiration of insects in air was measured using the same apparatus.

The table below shows results recorded by a student using this apparatus.

Organism	Distance moved by liquid in 15-minute intervals/mm				Mean rate of respiration/ mm min^{-1}
germinating seeds	7	6	5	6	0.4
germinating seeds in nitrogen gas	0	0	0	0	0
insects	12	11	13	12	

    (iv) Calculate the mean rate of respiration for the insects expressed as movement of liquid in millimetres per minute. Show your working. **(2)**

    (v) Explain modifications to the experiment that would allow a valid comparison to be made between the mean rates of respiration of the germinating seeds in air and the insects. **(3)**

(Total for question 2 = 18 marks)

3 Heart rate and body temperature are controlled in mammals.

    (a) (i) Name the tissue in the heart that controls resting heart rate. **(1)**

          ☐ **A** atrioventricular node

          ☐ **B** bundle of His

          ☐ **C** Purkyne tissue

          ☐ **D** sinoatrial node

      (ii) Name the heart activity that is shown by an ECG. **(1)**

          ☐ **A** blood pressure

          ☐ **B** cardiac output

          ☐ **C** electrical activity

          ☐ **D** stroke volume

      (iii) State the name of the part of the brain that controls heart rate. **(1)**

(Total for question 3 = 3 marks)

4 Nerve impulses are transmitted along the axon of a neurone.

    (a) The diagram shows the structure of a motor neurone.

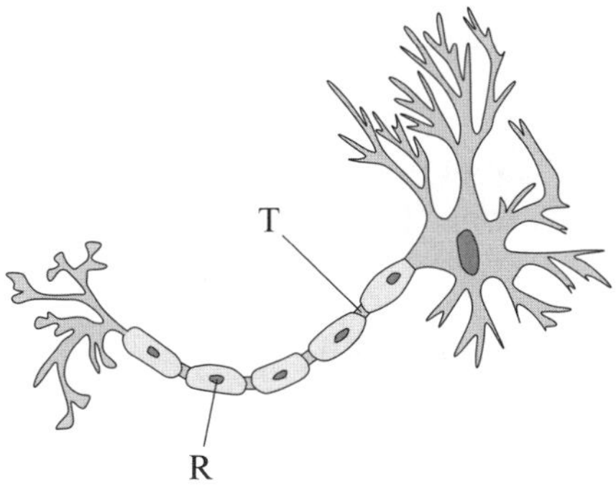

      (i) Name the part of the neurone labelled T. **(1)**

          ☐ **A** dendrite

          ☐ **B** node of Ranvier

          ☐ **C** Schwann cell

          ☐ **D** synapse

      (ii) Name the part of the neurone labelled R. **(1)**

          ☐ **A** synapse

          ☐ **B** nucleus of cell body

          ☐ **C** dendrite

          ☐ **D** nucleus of Schwann cell

(iii) The graph below shows changes in the membrane potential during the transmission of an impulse along the axon of a motor neurone.

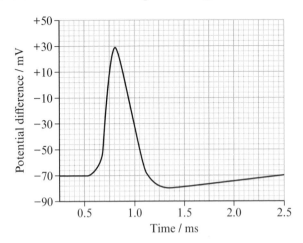

Place a cross in the box next to the description of the membrane potential at 0.75 ms on the graph. **(1)**

☐ **A** depolarised

☐ **B** hyperpolarised

☐ **C** polarised

☐ **D** repolarised

(iv) Explain how the structure of this motor neurone affects the speed of the impulse along the axon. **(2)**

(b) The skin of the golden poison frog (*Phyllobates terribilis*) produces a poison that affects sodium ion channels in the axon membrane of a neurone. The poison causes these channels to stay open.

(i) Explain the effect the poison has on the ability of a neurone to transmit impulses. **(4)**

(ii) Explain why the neurones of the golden poison frog are not affected if they come into contact with the poison. **(2)**

(Total for question 4 = 11 marks)

5 The diagram below shows a spirometer that can be used to measure lung volumes. A spirometer can also be used to measure the volume of oxygen a person uses.

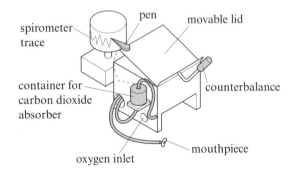

(a) A student used a spirometer to measure the volume of oxygen he used at rest and during exercise.

The spirometer trace below shows the results he obtained during exercise.

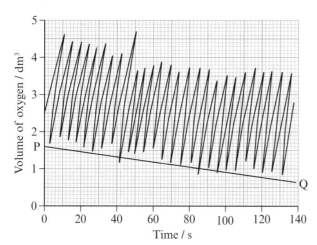

(i) The line P to Q slopes downwards because oxygen is being used.
Calculate the volume of oxygen used during one minute of exercise. **(3)**

(ii) The student had a body mass of 70 kg.
Calculate the rate of oxygen used by this student.
Show your working. **(3)**

(iii) Explain how this spirometer trace would differ from the one the student obtained at rest. **(3)**

(b) (i) The air the student exhaled passed through the carbon dioxide absorber in the spirometer.
Name a suitable carbon dioxide absorber. **(1)**

(ii) Explain why the spirometer trace would be different if the carbon dioxide had not been absorbed. **(2)**

(c) Explain how carbon dioxide is involved in the control of breathing rate during exercise **(5)**

(Total for question 5 = 17 marks)

6 Animals that are predators often show bursts of very fast movement. Their prey tend to be able to carry out sustained movement over longer periods of time. Close examination shows that the muscles of predator and prey show a different composition of fast and slow twitch fibres.

(a) (i) Compare and contrast fast and slow twitch muscle fibres. **(3)**

(ii) State whether predator or prey would show a higher proportion of slow twitch fibres. **(1)**

(iii) Discuss why predators show different proportions of fast and slow twitch muscle fibres from their prey. **(3)**

(b) During fast movement, lactate builds up in the muscles of the predator.
Explain what happens to this lactate after the chase has ended. **(3)**

(c) During the chase, the core body temperature of both predator and prey rises. Describe how changes in blood circulation help to return their core body temperatures to normal. **(4)**

(Total for question 6 = 14 marks)

7 Research on visual development in cats has led to knowledge of how information is processed in the visual cortex of the brain. The diagrams below show the growth of neurones in part of the visual cortex after normal visual development and when

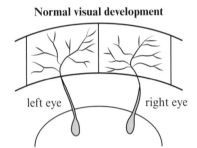

Normal visual development

left eye   right eye

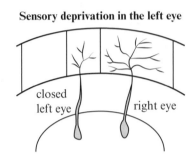

Sensory deprivation in the left eye

closed left eye   right eye

the left eye has been deprived of sensory information.

(a) Analyse the diagrams to describe differences in the visual cortex after sensory deprivation. **(3)**

(b) Explain how this type of experiment has provided evidence that shows the need for exposure to sensory information in normal visual development. **(2)**

(c) Describe other evidence that suggests that humans must be exposed to particular stimuli if they are to develop normal vision. **(2)**

(Total for question 7 = 7 marks)

8 IAA (auxin) is a plant growth substance.

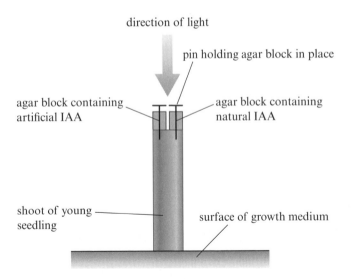

direction of light

pin holding agar block in place

agar block containing artificial IAA

agar block containing natural IAA

shoot of young seedling

surface of growth medium

(a) A student investigated the effect of natural IAA and artificial IAA on shoot growth. The diagram above shows how she set up her investigation.

(i) The student also set up a control. Describe a suitable control for this investigation. **(1)**

*(ii) After 48 hours, the student recorded her observations of the growth of the shoots. From her observations, she concluded that both natural and artificial IAA affected growth. She also concluded that the artificial IAA had a greater effect than the natural IAA. Analyse the information to explain how the IAA in the agar affected the growth of the shoot. **(6)**

(b) IAA is known to bind to transcription factors. Explain how IAA can stimulate cells to synthesise proteins. **(4)**

(Total for question 8 = 11 marks)

9 A sea slug uses a tube called a siphon to help with gas exchange. When the siphon is withdrawn into the body of the sea slug.

An investigation was carried out to compare the behaviour of a sea slug living in calm water with a sea slug living in rough water.

A squirt of seawater from a syringe was directed towards the siphon of each sea slug.

The time that the siphon remained withdrawn in the body was recorded.

Ten successive trials were carried out for each sea slug.

The graph below shows the results of this investigation.

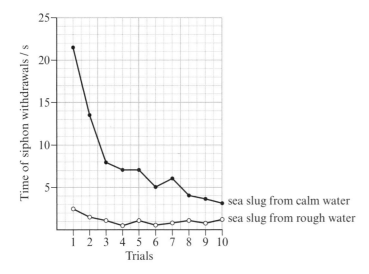

(a) Analyse the information in the graph to compare and contrast the behaviour of these two sea slugs. **(4)**

(b) Explain the advantage of this behaviour to the sea slug living in rough water. **(3)**

(c) Functional magnetic resonance imaging (fMRI) can be used to observe the activity of the human brain in response to repeated stimuli.

Explain how fMRI can be used to observe activity in the human brain in response to repeated stimuli. **(5)**

(Total for question 9 = 12 marks)

TOTAL FOR PAPER = 100 MARKS

Scan the QR code with your mobile phone or tablet using a QR reader to access more practice questions.

Answers

The answers provided here are possible responses. In some cases other answers may also be possible.

1. Why is transport needed?

1

Feature	Explanation
liquid at room temperature	Water molecules are joined to each other by 'sticky' hydrogen bonds. This stops them separating into separate molecules and thus becoming a gas.
polar solvent	Water is a dipole (the hydrogen end is slightly positive and the oxygen end is slightly negative) which pulls apart molecules held together by ionic bonds. Water surrounds the ions. Also, polar molecules, such as sugars and amino acids, become surrounded by water and dissolve.
high specific heat capacity	Much energy is needed to break the hydrogen bonds. This energy does not raise the temperature; a lot of heat is absorbed before it becomes hot.

(5)

2 (a) 6 cm^2 **(1)**
   (b) 33 cm^2 **(1)**
   (c) the thinner cuboid **(1)** larger surface area **(1)** shorter diffusion pathway / increased volume of gas can diffuse at the same time **(1)**

2. Blood vessels

1 (a) **(4)**

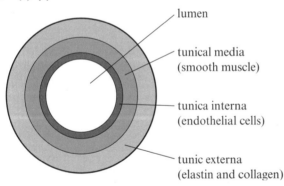

lumen

tunical media (smooth muscle)

tunica interna (endothelial cells)

tunic externa (elastin and collagen)

   (b) It has a thick wall to withstand blood under high pressure and a narrow lumen to maintain high pressure. The artery also has elastic fibres to allow it to stretch and then recoil to maintain pressure. Finally, there is a smooth lining to reduce friction. **(4)**

2 A 1 B 2 **(2)**

3. The heart

1 The heart has four chambers, the atria above (in humans) and the ventricles below.
The left and right sides are separated by a septum.
All walls of all chambers are composed of cardiac muscle.
The ventricle walls are thicker than the atria, the left ventricle being thicker than the right.
The atrioventricular valves separate the atria from the ventricles and semilunar valves are at the entrance to the arteries leaving the heart.
Tendons are found between the walls of the ventricles and the flaps of the AV valves.

The aorta leaves from the left ventricle and the pulmonary artery from the right ventricle.
The vena cava and pulmonary vein enter the atria.
The heart is supplied with coronary arteries.
The pacemaker is found in the right atrium and the atrioventricular node between the atria and ventricles.
Purkyne fibres radiate out from the apex of the heart. **(6)**

2 (a) to prevent blood flowing back again **(1)**
   (b) Valves are found in veins, in the pulmonary artery, the aorta and between atria and ventricles.
   They are present in veins because there is not a continuous high pressure pushing blood forward; blood is squeezed forward by body muscles and would fall back when these relaxed again.
   They are found between atria and ventricles because the blood would move back up into the atria when the powerful ventricle muscles contracted when pushing blood out of the heart.
   They are found in the pulmonary artery and the aorta to stop blood flowing back into the ventricles when they relax after pushing the blood out of the heart. **(4)**

3 Thicker muscle creates greater pressure to push blood all around the body. Too thick on the right side would lead to too high a pressure in the lungs leading to fluid accumulation there. **(2)**

4. The cardiac cycle

1 The cardiac cycle consists of three stages: atrial systole, ventricular systole and diastole. During atrial systole, the atria contract and the ventricles are relaxed. The atrioventricular valves are open. During ventricular systole, the semilunar valves open as oxygenated blood is forced out of the heart through the aorta to the body and through the pulmonary artery to the lungs. **(3)**

2 (a) It takes 0.8 s for one cycle, so in 60 seconds there will be 60 ÷ 0.8 beats. That is 75 bpm. **(2)**
   (b) (i) atrioventricular valve closes   0.13 s **(1)**
       (ii) aortic valve closes   0.32 s **(1)**
       (iii) atrioventricular valve opens   0.5 s **(1)**
   (c) During QRS the ventricles are depolarising, during T they are repolarising. **(2)**

5. Clots and atherosclerosis

1 A blood clot may form when a blood vessel wall becomes damaged. Cell fragments called platelets stick to the wall of the damaged blood vessel forming a plug. A series of chemical changes occur in the blood, resulting in prothrombin being converted into thrombin. Thrombin is an enzyme that catalyses the conversion of fibrinogen into long insoluble strands of fibrin. These strands form a mesh that traps cells to form the clot. **(6)**

2 Endothelial cells become damaged, perhaps due to inflammation. White blood cells accumulate in the damaged area. There is also a build-up of cholesterol and fibrous tissue and a plaque forms. This leads to a loss of elasticity of the artery and a narrowing of the lumen. **(4)**

3 The atheroma narrows arteries, which means there is reduced blood flow to cardiac muscle through the coronary artery. This means that the heart receives less oxygen and starts to respire anaerobically. A by-product of anaerobic respiration is lactic acid, which builds up and causes pain. **(3)**

6. Risk

1 high salt and high cholesterol **(1)**
(other answers include: high saturated fat, high trans-fat, high calories, high alcohol, low fibre, low NSP, low antioxidants, low vitamin C, low vitamin E)

2 anything that raises the chance of harm **(1)**

3

(3)

4 100% **(1)**

5 Due to an energy imbalance the person becomes obese leading to an increase in blood pressure. Obesity also leads to diabetes, which is a CVD risk factor.
The high-fat diet can lead to increased cholesterol levels and possible damage to endothelium of blood vessels. Atheromas may form leading to a loss of elasticity of arteries and a narrowing of their lumens. **(6)**

7. Correlation and causation

1 (a) There is very little difference between the cholesterol levels of the different ethnic groups. There appears to be no obvious correlation between cholesterol level and rates of high blood pressure. However, there is a difference in high blood pressure between ethnic groups – the black and African Americans have higher blood pressure than the white or Mexican Americans. **(3)**

(b) Although the data shows that high blood pressure is more common in black and African Americans, suggesting a genetic cause, there is no information about other variables. These include the mass, age and diet, together with smoking habits and levels of exercise undertaken. These are all risk factors for high blood pressure. In addition, there is no information on how many individuals were tested. **(4)**

2 Causation is when a change in one variable is responsible for a change in another variable. A correlation is a relationship between two variables such that a change in one of the variables is reflected by a change in the other variable but this may not necessarily be a direct cause of the change in the first variable. **(2)**

8. Studying the risks to health

1 Both studies have a control group.
A cohort study looks at past history to explain present observations and is therefore retrospective whereas a case-control study follows study groups into the future and is therefore prospective.
In a cohort study, the control group is selected for a match with the 'experimental' group whereas in a case-control study the control group is selected as those who do not develop the condition. **(3)**

2 It should have a large and representative sample. It should be long term and ethical. The measurement techniques used should be valid and the methods used reliable (for example, standardised procedures). **(4)**

3 The UK has a higher death rate from CHD than France by about seven times. However, in France the rate of smoking is higher by about 30%. French people also consume more alcohol, although fat intake is about the same. Since smoking and alcohol consumption are thought to be risk factors for CHD the incidence would be expected to be higher, not much lower. **(6)**

9. Energy budgets

1 $76 \div (1.70)^2 = 26.3$. This person is not obese. **(2)**

2 At a BMI of 25 the increased risk is 16%, whereas at 30 it is 26%. This an increased risk of 10%. Between 30 and 35 BMI it goes up another 19%. This suggests a non-linear relationship with increasing BMI causing an increasing rise in risk of death by CHD. **(3)**

3

Waist measurement	Hip measurement	Waist : hip ratio
97 cm	112 cm	0.87
61 cm	91 cm	0.67
40 inches	46 inches	0.87
52 cm	68 cm	0.76
127 cm	138 cm	0.92

(5)

10. Monosaccharides and disaccharides

1 A **(1)**

2 *two glucose molecules correctly drawn* **(1)**
indication that water is formed **(1)**
glycosidic bond correctly drawn **(1)**

3 The role of glucose is that it is the source of readily available energy for cellular energy-requiring processes. It is very soluble in water so it is easily transported in the blood. It is a small molecule so can enter cells quite easily.
Sucrose is a transport molecule in plants and is less reactive. So it is less likely to be broken down during transport. **(4)**

11. Carbohydrates – polysaccharides

1 Both glycogen and starch are made from α-glucose molecules. These glucose are joined by 1,4 and 1,6 glycosidic bonds in glycogen and amylopectin and 1,4 in amylose. These are easily broken by hydrolysis. Glycogen and starch both have low solubility so do not diffuse out of cells and have no osmotic effect. **(4)**

2

Statement	True	False
polymer of glucose	✓	
molecule contains α- and β-glucose		✓
glycosidic bonds present	✓	
molecules may have side branches		✓
molecules can form hydrogen bonds with adjacent molecules	✓	

(5)

3 Amylose is unbranched (because it has no 1,6 glycosidic bonds) so there are fewer places for it to be broken down, just either end of the molecule therefore it is broken down slowly to release glucose and thus energy. Amylopectin has many branches (due to 1,6 glycosidic bonds) and so there are more places where the glucose can be broken off leading to a faster release of glucose and therefore energy. **(2)**

4 Both are polymers of α-glucose and both are branched with 1,6 glycosidic bonds. Glycogen has more branches than amylopectin. **(3)**

12. Lipids

1 1 mark for COOH

(2)

2

showing a double bond anywhere along the chain **(1)**
showing the **two** missing hydrogens **(1)**

3 Total for concentration of fatty acid in $mg\,g^{-1}$ of breast milk in vegans is 982 of this, unsaturated is $657\,mg\,g^{-1}$ of breast milk.
Total for control group is $963\,mg\,g^{-1}$ of breast milk and, of this unsaturated is $466\,mg\,g^{-1}$ of breast milk
So percentage in vegans = (657 ÷ 982) × 100 = 66.9%
percentage in control group = (466 ÷ 963) × 100 = 48.4%
The proportion is higher in vegans than control group. **(3)**

13. Good cholesterol, bad cholesterol

1 The cholesterol levels in people with the mutation are not higher than in people without the mutation. LDL (cholesterol) levels in people with the mutation are not higher than in people without the mutation. HDL (cholesterol) levels in people with the mutation are not lower than people without the mutation. **(3)**

2 The deposition of cholesterol in artery walls may lead to atheroma formation. This in turn can narrow the lumen of coronary arteries and increase blood pressure, both leading to CHD. **(3)**

3 They are associated with the formation of atherosclerosis because they are formed from saturated fats, protein and cholesterol and bind to cell surface receptors, which can become saturated leaving the LDLs in the blood. **(3)**

14. Reducing the risk of CHD

1 More vitamin C is needed to decolourise the DCPIP as temperature increases. This shows that as temperature increases the vitamin C in the juice decreases. The effect is greatest between 20 and 30 °C. The concentration of vitamin C at each temperature is significantly different from that at the next temperature above and below since the SDs do not overlap. **(4)**

2

Decrease	Increase
saturated fat	unsaturated fats
salt	fruits
cholesterol	oily fish
	non-starchy polysaccharides

(2)

15. Medical treatments for CVD

1 (a) 14.8 − 8.9 = 5.9; (5.9 ÷ 14.8) × 100% = 39.9% **(2)**
　(b) (i) so that the validity of the data can be measured **(1)**
　　　(ii) to confirm the drug was having the effect **(1)**

2 (a) Platelet inhibitory drugs prevent platelets becoming sticky and therefore prevent the formation of a blood clot leading to a stroke. Antihypertensives reduce blood pressure by reducing heart rate. **(4)**
　(b) Beta blockers have links with increased incidence of diabetes. Aspirin can cause severe stomach bleeding. **(2)**

16. *Daphnia* heart rate

1 (a) When one variable (caffeine concentration) changes there is also a change in an accompanying variable (*Daphnia* heart rate). **(1)**
　(b) Since there are seven means (see part a) the critical value for significance is 0.79. The calculated value of 1 is greater than the critical value so it can be concluded that there is a positive correlation between the concentration of caffeine and the heart rate of the *Daphnia*. Low concentrations have a large effect; higher concentrations produce a smaller increase. Caffeine is a known stimulant and it could be suggested that it has an effect on the *Daphnia*'s nervous system. **(4)**
　(c) Because *Daphnia* is an invertebrate it has a simple nervous system and thus a reduced awareness of pain. **(2)**

17. Exam skills

1 (a) *Daphnia* are transparent so the heart is easily visible without an invasive procedure. They are also cheap and abundant, together with being well studied and easy to manipulate. **(2)**

　(b) *Daphnia* have a simple nervous system and probably reduced awareness of pain. They are bred for fish food and therefore are destined to die anyway. **(2)**

(c)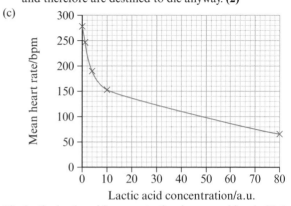

(5)

(d) As the lactic acid concentration increases blood pH the heart rate of *Daphnia* lowers/decreases. This is likely to be due to the effect of pH of enzymes, which are known to have optimum value for action. Increased lactic acid concentration lowers the blood pH of the *Daphnia*. The beating of a heart involves a number of metabolic processes that are enzyme catalysed. **(6)**

18. Gas exchange

1 (a)

Length of side/ cm	Area of surface of cube/ cm²	Volume of cube/ cm³	SA/V	Time for whole block to become coloured/ seconds
13	1014	2197	0.46	380
10	600	1000	0.60	300
7	294	343	0.86	100
5	150	125	1.20	53
3	54	27	2.00	20

(5)

(b)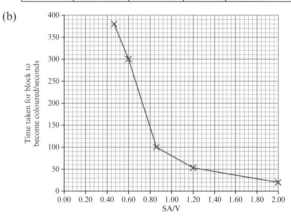

(4)

(c) The graph shows that as the surface area to volume ratio increases, the time decreases for the block to become coloured throughout decreases. Small organisms have a large surface area to volume ratio and the graph shows that such a situation leads to substances reaching the centre of these small objects, blocks or animals, much more quickly than the larger ones.
So, smaller animals do not need a specially adapted gas exchange surface such as gills or lungs. **(4)**

19. The cell surface membrane

1 The fatty acid 'tails' of the phospholipids are hydrophobic. This means they orientate themselves away from water, which is found inside and outside cells. The phosphate heads are hydrophilic and can interact with water, so they orientate themselves towards aqueous solutions. A bi-layer is formed with the phosphate heads on both the outer sides and the fatty acid 'tails' on the inside. **(3)**

2 Size of image (that is, distance from A to B) = 6 mm
so real size of object = 6 ÷ 1 000 000 mm = 0.000006 mm.
That is, 6×10^{-9} m or 6 nm. **(3)**

20. Passive transport

1 There is a clear positive relationship between solubility in oil and ability to cross the membrane. This supports the part of the fluid mosaic model that says that the membrane is made of a phospholipid bilayer, part of which is hydrophobic. There is no support from these data for the aspect of the model which suggests that the bilayer is studded with proteins, some of which form channels. There is no obvious correlation between size and permeability. **(4)**

2

Process	Requires energy from respiration (ATP)	Requires a concentration gradient
passive diffusion	incorrect	correct
facilitated diffusion	incorrect	correct
osmosis	incorrect	correct
active transport	correct	incorrect

(4)

21. Active transport, endocytosis and exocytosis

1 Both processes involve the use of vesicles to move contents in bulk transport. Both processes also require energy in the form of ATP.
Endocytosis moves substances into the cell but exocytosis transports substances out of the cell. **(3)**

2 (a) **A**
(b) **B**
(c) **C**
In b the plateau is due to the fact that the carrier protein is saturated and carrying as many molecules as it can in a unit of time. This plateau is reached at a higher concentration in c due to the presence of the competitor.
Marks are 1 for the three correct answers and 4 for the explanations.

22. Practical on membrane structure

1 (a) blue **(1)**
(b) biotic variables: *any two from:* surface area/volume (of beetroot), part of beetroot used, age of beetroot used, variety of beetroot used; abiotic variables: *any two from:* storage conditions, temperature, wavelength filter **(4)**
(c) At 0% ethanol there was some discolouration. This may have been due to cells or membranes being damaged by cutting up the beetroot into pieces. As a result, pigment could leak out of vacuoles (cells). **(2)**
(d) Increased ethanol concentrations increase intensity of discolouration and therefore disruption of the membrane. Ethanol is a non-polar/organic solvent, which dissolves lipids. Increase in ethanol causes the solution to be less polar and changes the orientation of the phospholipids as this depends on the polar nature of the water around them. **(3)**

23. The structure of DNA and RNA

1 **A (1)**

2

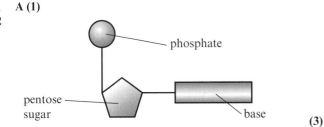

phosphate

pentose sugar

base

(3)

3 The phosphate group is attached to the deoxyribose sugar at carbon atom number five by a condensation reaction and a base is attached to the deoxyribose sugar at carbon atom number one by a condensation reaction. **(3)**

24. Protein synthesis – transcription

1 X is DNA; Y is mRNA; Z is protein **(3)**
2 (a) uracil and cytosine **(1)**
(b) C T A G C A C **(1)**
(c) 200 × 3 = 600 bases + 3 for the stop codon = 603 ÷ 50 = 12.06 seconds **(3)**

25. Translation and the genetic code

1 Both a gene and a codon are a sequence of bases. However, a gene is many bases and codes a polypeptide chain while a codon is a sequence of three bases which codes for an amino acid. **(3)**
2 Degeneracy is where amino acids are coded for by more than one triplet codon. This means that some mutations may have no effect on the sequence of amino acids. **(2)**
3 Step 1 **C**
Step 2 **D**
Step 3 **A**
Step 4 **B (2)**
4 Transcription is when one of the strands of DNA is used as a template for the synthesis of mRNA. Ribonucleotides are paired with their complements on the template strand, and uracil instead of thymine pairs with adenine. RNA polymerase then joins up the ribonucleotides.
Translation is when a protein is synthesised when tRNA's bases pair with the complementary codon on the mRNA, bringing designated amino acids with them. These amino acids are then bonded to each other to create a polypeptide chain. **(5)**

26. Amino acids and polypeptides

1 B **(1)**

2

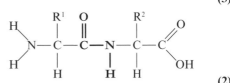

(3)

3 (a)

$$H - N - C - C - N - C - C - OH$$

(2)

(b) dipeptide **(1)**
(c) No, because they have different R groups; the R group differentiates one amino acid from another. **(1)**
4 C **(1)**

27. Folding of proteins

1 (a) Primary structure is the sequence of amino acids in the chain, for example, in this case, Lys-Glu-The-Ala-Ala-Ala etc.
Secondary structure is the part of the structure stabilised by H bonds between the backbone of the molecule. In this case, there would be hydrogen bonds in the region around D.
Tertiary structure is the structure which is stabilised by interactions between R side chains. Here, the Cys-Cys held together by disulfide bridges would be an example. **(6)**
(b) A is carboxyl; C is amino **(2)**
(c) B is a disulfide bridge. It is a covalent bond. **(2)**
(d) A polypeptide chain will need to be folded to create a globular shape, which is precise for its function as an enzyme. It may have to associate with other polypeptides if the enzyme has a quaternary structure. **(2)**

28. Haemoglobin and collagen

1 Fibrous proteins are long chains whereas globular proteins are spherical. Both consist of repeated chains of amino acids. Fibrous proteins are long, unfolded chains whereas globular are spherical. Globular proteins are folded whereas fibrous proteins are not. Finally, globular are soluble in water but fibrous are not. **(3)**

2   Collagen is made up of amino acids in a repeating sequence
    of three, one of which is always glycine. The other two are
    commonly proline and hydroxyproline. This primary structure
    is twisted into a helix held together by H bonds between parts
    of the molecule's backbone. Three such helices are twisted
    together to form a triple superhelix held together by H bonds
    between the R groups of the amino acids. **(5)**

3   It is a globular protein so is soluble in water. It has iron-
    containing haem groups that allow it to bind to oxygen. It has
    four haem groups, which means it can carry a large amount of
    oxygen. Its shape changes when it binds to one oxygen group
    so that it binds to the next one more easily and so on to the
    fourth. This cooperative binding property means it is good at
    binding with oxygen when there is a high level of oxygen, but
    not very good when oxygen levels are low. **(5)**

29. Enzymes

1   **C (1)**

2   (a) One of the features of enzymes is that they are specific
        to their substrate. Caffeine is the substrate in this case.
        Because the products are different shapes, P450 must
        consist of at least three enzymes with different active
        sites that will join with caffeine but which then lead to a
        different reaction and a different product. **(3)**

    (b) theobromine 7.14 mg; theophylline 2.38 mg
        $84 \div 12 = 7$, so $50 \div 7 = 7.14$ mg of theobromine would be
        produced
        4% theophylline would be one-third of this, $7.14 \div 3 =$
        2.38 mg **(4)**

30. Activation energy and catalysts

1   **B (1)**

2   Enzymes can put a strain on bonds in the substrate to break
    them. They may also provide a more favourable pH in the
    active site. Finally, they can bring reactants close together in
    the active site so bonds are easier to form. **(3)**

3   A is the energy rise in the substrate needed to get the reaction
    to happen when an enzyme is present. It is much lower than B.
    B is the energy rise in the substrate needed to get the reaction
    to happen without an enzyme, for example, by heating it.
    C is the difference in the energy possessed by the substrate minus
    the energy possessed by the products; this is given out as heat. **(3)**

31. Reaction rates

1   Enzyme reactions are very fast and the substrate (in this case,
    hydrogen peroxide) is rapidly used up and so the rate changes.
    This means that measuring something to an endpoint, as is
    suggested in this experiment, will not tell us anything useful about
    the initial rate of the reaction. It will only tell us the average rate
    over the time that the reaction is allowed to run for. **(4)**

2   In experiment 1, as the protein is broken down the cloudiness
    reduces and the absorbance reading falls. In experiment 2,
    the initially clear solution becomes cloudy as starch, which is
    insoluble in water, is formed. **(3)**

32. Initial rates of reaction

1   (a) absorbance reading **(1)**

    (b) Variables are things that might change the initial rate,
        other than temperature. The important one here would
        be pH, which would be achieved using a buffer solution.
        Other conditions that would be relevant are enzyme and
        substrate concentration, which are already maintained in
        the protocol. **(3)**

    (c) A graph is plotted of time on the *x*-axis and the
        absorbance reading on the *y*-axis. The gradient of the
        straight line portion of the graph is then calculated. This
        will be the initial rate in absorbance units per second. **(3)**

2   In a thermostatically-controlled water bath that is
    regularly monitored by taking temperature readings with a
    thermometer. **(2)**

33. Exam skills

1   (a) The enzyme increases the rate of reaction. The rate with
        enzyme is higher than the rate if no enzyme is present at
        all values of substrate concentration [S]. **(3)**

    (b) (i) ester **(1)**
        (ii) glycerol and fatty acids **(2)**
        (iii) The pH would be reduced due to the production of
             fatty acids, which would ionise to give rise to protons.
             (3)

34. DNA replication

1   Step 1 is R; Step 2 is S; Step 3 is P; Step 4 is Q **(2)**

2   **C (1)**

3   (a) deoxyribose **(1)**; (b) phosphate **(1)**; (c) phosphodiester bond
    (1); (d) nucleotide **(1)**; (e) thymine **(1)**; (f) hydrogen bond **(1)**

35. Evidence for DNA replication

1   (a) New DNA is synthesised which contains the original
        strand and the new strand. **(2)**

    (b)

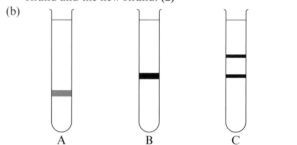

        A   B   C   **(2)**

    (c) At the time of Meselson and Stahl there were three
        theories of DNA replication. Most people accepted the
        semi-conservative hypothesis, although there was no
        direct evidence to support it. The other theories were
        the conservative and fragmentary. Meselson and Stahl's
        experiments showed that the first generation of DNA
        was of a mixed mass. Conservative replication would
        produce light and heavy forms so this was eliminated as
        a possibility. Fragmentary would give all mixed after two
        generations, whereas semi-conservative would give light
        and mixed, which is what they found. **(6)**

36. Mutation

1   A gene is a sequence of bases that codes for a sequence of
    amino acids in a protein.
    A mutation is a change in that sequence, so this may result in a
    different amino acid being inserted into the polypeptide chain.
    This may change the shape of the protein. Since enzyme
    function depends on protein shape this differently-shaped
    protein would not function as an enzyme. **(4)**

2   **D (1)**

3   (a) If the mutation caused the mutated codon to become a
        stop codon rather than coding for an amino acid, then
        protein synthesis would stop at this point. **(2)**

    (b) If the mutation caused the mutated codon to code for a
        different amino acid, a protein would be formed but it
        would fold imperfectly and thus not do its intended job. **(2)**

37. Classical genetics

1   Let the red allele be represented by R and the white allele by W.

parent's phenotype	red	white		
parent's genotype	RR	WW		
parent's gametes	R	W		
offspring genotype		RW		
crossed	RW	×	RW	
offspring gametes	R W		R W	
their offspring genotype	RR	RW	WR	WW
phenotype	red	pink	pink	white
ratio:	1 red : 2 pink : 1 white **(4)**			

2   Since 8 and 9 do not have galactosaemia but some of their
    children do, they must be heterozygous carriers. This means
    galactosaemia is caused by a recessive allele. When 8 and 9

have another child they will produce gametes N and n in ratio 1 : 1. Therefore, there will be a 0.5 × 0.5 (= 0.25) chance of the child being nn, that is, having galactosaemia. **(4)**

38. Cystic fibrosis symptoms

1 Gas exchange depends on diffusion, and rate of diffusion depends on, amongst other things, the thickness of the diffusion path. With a build up of sticky mucus the path becomes thicker and diffusion slows down. **(2)**
2 Thick mucus accumulates in the reproductive system so sperm reaches the egg much less efficiently. If the egg is fertilised, implantation can be impaired. **(2)**
3 Physiotherapy can help to loosen the mucus, which improves airflow into and out of lungs, reducing the risk of infection. Digestive enzyme supplements can be given and these can substitute for those that cannot be obtained from the pancreas as the pancreatic duct is blocked.
Antibiotics destroy bacteria trapped in the mucus, reducing infection rates. **(3)**
4 The trapped enzymes inside the pancreas can damage the insulin-producing cells. This reduces insulin concentration so regulation of blood glucose is impaired. **(2)**

39. Genetic screening

1 There is a 1–2% risk of miscarriage so the couple may lose the baby. They might also be concerned about what they should do if the result was positive. On the other hand, they would be prepared if they knew that the baby was going to have CF. **(2)**
2 CVS can be carried out earlier than amniocentesis so if an abortion is needed it can be carried out earlier too. An earlier abortion is preferable to a later one since the procedure will be less invasive/traumatic for the parents. **(2)**
3 The DNA of the embryo is analysed for the presence of the faulty CFTR allele. Implantation is carried out using an embryo in which it is known the disease does not occur. **(2)**
4 One ethical issue is who has the right to decide if tests should be performed and there may be disagreements over the next step. The success rate is low at 30% so the couple may suffer disappointment and this also makes the procedure costly, which is a problematic social issue if it is on the NHS. Another ethical issue is what to do with the spare embryos. **(4)**

40. Exam skills

1 (a) A is adenine; C is cytosine; G is guanine; T is thymine **(1)**
  (b) Triplet means the bases on three nucleotides code for each amino acid. In this case, twelve nucleotides code for four amino acids. For example, AAT codes for leucine.
   Non-overlapping means that each triplet is discrete. So, in this case, AAT, AAC, CAG and TTT gives four separate codes.
   Degenerate means that more than one code can be used for a particular amino acid. In this case, AAT and AAC both code for leucine. **(3)**
  (c) B **(1)**
  (d) A strand of mRNA with sequence UUA UUG GUC AAA would be formed. This would bind to a ribosome that is involved in protein synthesis. tRNA molecules attached to one specific amino acid would then bind to the codon on the mRNA which is complementary to their anticodon. For example, in this case, a tRNA carrying the amino acid valine would have anticodon GUC and would pair with the CAG region of the mRNA. The two are held together by hydrogen bonds. This would also happen on the next codon with another tRNA/amino acid pairing. Finally, peptide bonds are formed between the two amino acids on the tRNAs by enzymes associated with the ribosome to make part of a polypeptide chain. **(5)**

41. Exam skills

1 (a) Because at this point the substrate concentration limits the rate. This means that some active sites are not occupied for some of the time. **(2)**

(b) In this case, the independent variable is the enzyme concentration and the dependent variable is the initial rate of reaction. A suitable range of concentrations of enzyme would have to be chosen; this would be done best by carrying out a preliminary experiment in order to see how fast the substrate was broken down. The volume of enzyme and substrate should remain unchanged in each run. Non-experimental variables, particularly temperature and pH, should be controlled: the former by using a thermostatically-controlled water bath and the latter by using a buffer. Depending on the enzyme-substrate system used, the time course of the reaction would need to be followed for each enzyme concentration. **(5)**

42. Prokaryotes

1 (a)
(note 2 marks for ribosome: 1 for 70S and 1 for ribosome)

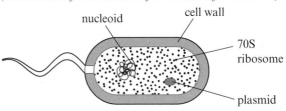

(5)

2 In order to be able to see the interior structures of bacterial cells they have to be viewed through an electron microscope. At this point they have undergone extensive preparation using various chemicals and they are dead. For these reasons most scientists agree that mesosomes are produced by this procedure and regard them as artefacts. **(2)**
3 They contain genes that aid the bacterium's survival, such as antibiotic resistance or toxin-producing genes. **(3)**
4 tick in 'circular DNA'
tick in 'small (70S) ribosome' **(2)**

43. Eukaryotes

1

Description	Name	Function
a large organelle with a double envelope with pores through it	nucleus	stores DNA
a branching series of channels studded with small, roughly spherical structures	rough endoplasmic reticulum	synthesis and transport of protein
quite large oval organelles with folded membranes inside	mitochondria	the site of respiration
a pair of cylindrical structures at right angles to each other	centrioles	make the spindle fibre in cell division

(4)

2

Animal cell only	Plant cell only	Animal and plant	Bacteria only	All three cell types
centrioles		mitochondria smooth endoplasmic reticulum (sER) DNA in a nucleus		ribosomes cell surface membrane

(3)

3 The organelles not found in animals are chloroplasts where photosynthesis occurs, and a vacuole where water and minerals are stored. **(4)**

44. Electron micrographs

1 (a) A is nucleus; B is rough endoplasmic reticulum; C is mitochondrion; D is nucleolus **(2)**
(b) ribosomes **(1)**
(c) smooth endoplasmic reticulum **(1)**
(d)

(1)

45. Protein folding, modification and packaging

1 The protein is released from ribosomes and enters the lumen of the rER.
In here, it folds and becomes packaged into rER vesicles. These move to Golgi apparatus where they fuse with the Golgi membrane. The protein is modified by the addition of carbohydrate to make a glycoprotein. They then become packaged into secretory vesicles where they become part of the vesicle membrane, which then fuses with the cell membrane, incorporating the glycoprotein into it. **(4)**

2

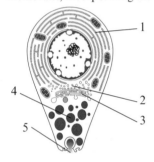

(5)

46. Sperm and eggs

1 A site containing an enzyme which digests the zona pellucida   **C**
A site with a haploid number of chromosomes   **B**
A site containing mitochondria   **G**
The zona pellucida   **F (4)**

2

Feature	Egg only	Sperm only	Sperm and egg	Neither sperm nor egg
mitochondria			✗	
DNA			✗	
cortical granules	✗			
membrane			✗	
cell wall				✗
diploid nucleus				✗
mid-piece		✗		

(7)

47. Genes and chromosomes

1 **C (1)**

2 (a)

VvBb × VvBb
/
VvBb
/

GAMETES	VB	vb
VB	VVBB	VvBb
vb	VvBb	vvbb

VvBb ⟨

giving
VVBB = normal wing, normal body
VvBb = normal wing, normal body
vVbB = normal wing, normal body
vvbb = vestigial wing, black body,
so ratio is
normal wing, normal body 3 : vestigial wing, black body 1 **(4)**

(b) Because V and B could be separated and so could v and b, there would be fewer normal wing, normal body and vestigial wing, black body, and some of phenotypes normal wing, black body and some vestigial wing, normal body. The exact ratio of the four phenotypes would depend on how far apart the genes are on the chromosome and thus how often crossing over occurred. **(3)**

48. Meiosis

1 It is cell A because the chromosomes are lined up in their homologous pairs and there is evidence of crossing over; neither of these is true of B. **(3)**

2

(1)

3 If two genes are linked then we would expect the offspring from two individuals, heterozygous for both genes, to be either the phenotype of one or other of the original parents of the heterozygotes. Mendel assumed that genes are not linked and predicted nine individuals like the double dominant parent; one individual like the double recessive parent; three individuals showing the dominant phenotype and the recessive; and three individuals showing the recessive phenotype and the dominant. With crossing over, depending on how far apart the loci are on the chromosome, there will be fewer double dominant and double recessive individuals than Mendel would have predicted but more of the dominant recessive and recessive dominant individuals. However, unless the loci are at the extreme ends of the chromosome, it is unlikely that Mendel's prediction would be met in linked genes. **(4)**

49. The cell cycle

1 The chromosomes condense and therefore become visible as two chromatids. The nuclear membrane breaks down. Centrioles move to opposite ends of the cell and form spindles out of microtubules between the centrioles. The chromosomes line up on the equator attached to the centrioles by spindle fibres from their centromeres. The centromeres now split, giving rise to two separate chromatids which are pulled apart by contraction of the spindle fibres. When they reach the poles of the cell, anaphase is over. **(4)**

2 (a) A is metaphase; B is anaphase **(2)**
(b) This cell is in interphase and it is possible to tell this because the chromosomes are not visible. **(2)**

50. Mitosis

1 (a)

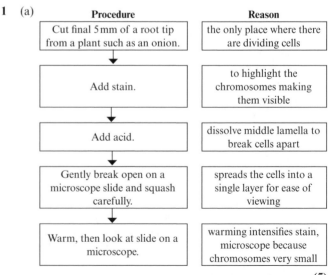

Procedure	Reason
Cut final 5 mm of a root tip from a plant such as an onion.	the only place where there are dividing cells
Add stain.	to highlight the chromosomes making them visible
Add acid.	dissolve middle lamella to break cells apart
Gently break open on a microscope slide and squash carefully.	spreads the cells into a single layer for ease of viewing
Warm, then look at slide on a microscope.	warming intensifies stain, microscope because chromosomes very small

(5)

(b) ethanoic (acetic) orcein (*Other suitable stains are available*) **(1)**

2 (a) 16 **(1)**
(b) mitosis **(1)**

3 Conditions in summer are stable and suitable for greenfly so no need to generate variation. In winter and the following spring, conditions might change so varied eggs are laid. Asexual reproduction is also a faster process so the animals can exploit the food quickly. **(2)**

51. Exam skills

1 DNA is synthesised in the S phase of the cycle. For the bud to have its own nucleus, mitosis is needed. Due to these two processes the bud will have the same genes as the original cell. Cytokinesis at the end of the cycle causes the bud to form. However, before this can happen, cytoplasm has to increase. This will include the synthesis of more organelles during interphase. **(3)**

2 In both cases, each chromosome consists of a pair of sister chromatids. In the first division of meiosis, homologous chromosomes pair but this does not happen in the second division. In the first division of meiosis, chiasmata are formed and this results in crossing over. This does not happen in the second division. Finally, in the first division, chromosomes assort independently but they do not do so in the second division. **(4)**

3 Crossing over recombines existing alleles, whereas mutation produces new alleles.
Mutations happen in any cell division; crossing over only occurs in meiosis.
Mutations are more likely to be non-viable. **(3)**

52. Stem cells and cell specialisation

1 In cell C all genes are potentially active as no genes are switched off, whereas cells A and B both have some genes switched off. Cell C can differentiate to become all cell types, whereas differentiation has occurred in cells A and B. **(3)**

2 (a) Stem cells are totipotent and can differentiate into specialised cells. These can replace damaged cells in the spinal cord. They are capable of continuous division due to no Hayflick limit. **(2)**

(b) The cells are genetically identical to the rest of the dog's cells so there is no rejection caused by the immune response. **(2)**

(c) This is used as a control to compare with the stem cell treatment. A control is needed for comparison. **(2)**

53. Gene expression

1 B **(1)**

2 Acetylation affects the histone proteins in which DNA is wrapped. When they are acetylated, wrapping is loosened, which exposes the genes so that they can be transcribed into active mRNA. Methylation affects DNA itself. When a methyl group is attached to DNA, transcription is prevented and no active RNA can be used. So, methylation switches genes off, demethylation switches genes on. **(4)**

3 A tissue is a group of identical or similar cells all performing the same function. An organ is a collection of tissues that work together to carry out one function or a range of functions. **(4)**

4 The promoter enables a gene to be transcribed.
The operator binds with the repressor.
The regulator codes for repressor protein.
The structural genes are the genes that are regulated by the operon. **(4)**

54. Nature and nurture

1 Non-identical twins are genetically different. Identical twins are genetically the same, meaning that any difference in their height and mass must be due to factors in the environment. There is a greater difference in height and mass for non-identical twins, suggesting these traits are mainly due to genetic effects. **(3)**

2 The error rate is the same for both in poor conditions and it is improved for both in enriched conditions. In enriched conditions, there is little difference between the two types. The maze-dull rats made fewer errors in an enriched environment compared with the normal one, showing that this is due to the environment. In a poor environment, the maze-bright did no better than the maze-dull but the differences shown in normal environments must be due to genes. The maze-bright that showed no improvement between normal and enriched environments had reached their full genetic potential in normal environments. **(4)**

55. Continuous variation

1 There are multiple genes which affect a single characteristic. The genes are at more than one locus on chromosomes and interact to determine the phenotype. **(2)**

2 (a) Each gene is double dominant so adds $2 \times 3\,cm$ to the height. There are four such genes, so this is $4 \times 6\,cm$.
LL adds 40 cm, so AABBCCDDLL is $(4 \times 6\,cm)$ + 40 cm = 64 cm
None of the genes a to d add anything, so the height is just the 40 cm from the LL, so aabbccddLL is $(0 \times 6\,cm)$ + 40 cm = 40 cm **(2)**

(b)

aaBbCcDDLL × AaBbCcDDLL

	ABCDL	ABcDL	AbCDL	AbcDL	aBCDL	abCDL	aBcDL	abcDL
aBCDL	AaBBCCDDLL 58 cm	AaBBCcDDLL 58 cm	AaBbCCDDLL 58 cm	AaBbCcDDLL 55 cm	aaBBCCDDLL 58 cm	aaBbCCDDLL 55 cm	aaBBCcDDLL 55 cm	aaBbCcDDLL 52 cm
abCDL	AaBbCCDDLL 58 cm	AaBbCcDDLL 55 cm	AabbCCDDLL 55 cm	AabbCcDDLL 52 cm	aaBbCCDDLL 55 cm	aabbCCDDLL 52 cm	aaBbCcDDLL 52 cm	aabbCcDDLL 49 cm
aBcDL	AaBBCcDDLL 58 cm	AaBBccDDLL 55 cm	AaBbCcDDLL 55 cm	AaBbccDDLL 52 cm	aaBBCcDDLL 52 cm	aaBbCcDDLL 52 cm	aaBBccDDLL 52 cm	aaBbccDDLL 49 cm
abcDL	AaBbCcDDLL 55 cm	AaBbccDDLL 52 cm	AabbCcDDLL 52 cm	AabbccDDLL 49 cm	aaBbCcDDLL 52 cm	aabbCcDDLL 49 cm	aaBbccDDLL 49 cm	aabbccDDLL 46 cm

(4)

56. Epigenetics

1 Epigenetics means 'above genetics'. It says that cells retain a 'memory' of the way they expressed their genes. There is some evidence that this memory can be passed on to daughter cells. It is a change in phenotype without a change in genotype. **(3)**

2 C **(1)**

3 Because daughter cells are just the product of mitosis so could be produced in growth or repair. To be passed from one generation to the next we would need evidence that they are passed on to daughter cells that arise from meiosis. **(3)**

4 They are proteins in cell nuclei that package the DNA into units called nucleosomes. They are also involved in the regulation of gene expression. **(3)**

5 Mutations are caused by changes to DNA base sequences. Epigenetic changes do not alter DNA base sequences. The DNA retains correct information on how to produce a polypeptide with epigenetic changes, whereas mutation may lead to different polypeptides being produced. Epigenetic changes alter the degree to which a gene is expressed. **(3)**

57. Biodiversity

1 The sharing strategy leads to more species of birds, that is, more diversity. Therefore, it is better to light log all the land rather than to intensive log some of the land leaving some land unlogged as 'set-a-side'. **(3)**

2 Three plots should be marked out in the same general area. All should be fenced to stop sheep from entering or leaving. One plot should be left untouched, one should be mown and one should have sheep introduced at suitable times. After a suitable period of time (say one year), sampling of the plants should be carried out. Quadrats of suitable sizes (predetermined) should be placed randomly in both plots. Plant counts should be carried out to give data about both the number of different species and the number of individuals of each species. The results can then be compared using a diversity index calculation based on the formula:

$$D = \frac{N(N-1)}{\sum n(n-1)}$$

The figures obtained from this will allow comparison of the species diversity of the three sites. The higher D, the more diverse the habitat. **(6)**

58. Adaptation to niches

1

Adaptation	Behavioural	Physiological	Anatomical
production of formic acid as an alarm signal	✓	✓	
acting in a group to carry heavy prey to the nest	✓			**(2)**

2 It is likely that the 1% which are not killed by the product are in some way resistant to it. This means that they will have access to any resources (food etc.) unhindered by their competitors, which have been killed by the product. They will quickly reproduce (being bacteria they may double in number as rapidly as once every 30 minutes or so) and pose a bigger threat than before the application of the product. **(3)**

3 The probability of bacteria becoming resistant to a particular antibiotic increases if a person does not finish the course because then too many survive and the remaining bacteria may mutate as they reproduce and develop resistance. **(3)**

4 Grey outcompeted red in lowland where it can get oak seeds. On higher ground, the grey is too big to survive on pine seeds. So the present distribution will probably be red in high pine woods and grey in lower oak ones. **(3)**

59. Evolution and speciation

1 (a) frequency (f) of R is $q = 0.1$
   $p + q = 1$, therefore $p = 0.9$
   frequency of RW is given by $2pq = 2 \times 0.1 \times 0.9 = 0.18 = 18\%$ **(3)**

   (b) The data tell us that now fRW = 0.32, so $2pq = 0.32$
   fRR = 0.04, so $q^2 = 0.04$, therefore $q = \sqrt{0.04} = 0.2$, thus $p = 0.8$
   so the frequency of the allele for red (R) has increased from 0.1 to 0.2, and that for white has gone down from 0.9 to 0.8. This would be due to one of the assumptions of the Hardy-Weinberg equilibrium not being met. For example, selection by pollinators may have favoured red and pink flowers. **(4)**

2 The original population may have spread into a wider diversity of habitats. Mutations may have occurred, leading to a diversity of flowering times. The populations may have become reproductively isolated, restricting gene flow. This would have been particularly apparent at the extremes of the population. In each of the new regions it is likely that different selection pressures would act on the populations, and so plants in these two areas would become adapted in different ways. This would lead to differences between the gene pools of the new populations. **(6)**

60. The classification of living things

1 The organism belongs to the Bacteria because the first amino acid in protein synthesis in bacteria is always formylmethionine; they have a peptidoglycan cell wall; and they are sensitive to antibiotics. None of these features is found in Archaea or Eukaryota. **(3)**

2 The organisms need to be placed in the optimum environment for them to be able to reproduce. There will have to be a mix of mature males from one habitat and mature females from the other habitat (and vice versa in a separate mating group) in an appropriate number according to their observed mating behaviour. They will need to produce offspring that survive and in due time mature and themselves reproduce to produce viable offspring. **(6)**

61. The validation of scientific ideas

1 **B**

2 The research is checked for its:
   validity – are its conclusions based on good methods and is the data reliable
   significance – does it make a useful addition to the existing body of scientific knowledge
   originality – has someone else already done the same work. This is done by someone respected in the field. **(4)**

3 He could have delivered his findings in a paper read at a scientific conference, or he could have had a poster in the poster display or in the media. **(3)**

4 The DNA molecule is the same in all organisms, supporting Darwin's idea of descent from a common ancestor. Estimates of the speed of mutation in DNA have shown that species have evolved over vast periods of time, just as Darwin thought. **(2)**

62. Plant cells

1 (a) **D (1)**
   (b) **C (1)**
   (c) **A (1)**
   (d **D (1)**
   (e) **C (1)**

2 plasmodesmata and pits **(1)**

63. Cellulose and cell walls

1 (a) Both are made up of glucose molecules and have 1,4 glycosidic bonds. Starch is α-glucose whereas cellulose is β-glucose. Furthermore, starch is composed of more than one type of molecule, amylose and amylopectin. In amylopectin there are branches due to 1,6 bonds but cellulose has straight chains. Finally, all the monomers in starch are orientated in the same direction every other one is inverted in cellulose **(4)**

   (b) They give the molecule tensile strength through a parallel arrangement of the separate straight chain molecules. These are held together by hydrogen bonding **(2)**

2 They hold cellulose molecules together by forming cross-links between them to form microfibrils. Although each one is quite weak, hydrogen bonds are collectively very strong because there are many of them. **(3)**

3 lignin holds the microfibrils together and keeps them parallel **(3)**

64. Transport and support

1 light because of lack of cytoplasm/hollow/no cell contents **(1)**
   strong due to presence of lignin/thickened cell walls **(1)**
   waterproof due to lignin **(1)**
2 (a) longitudinal **(1)**
   (b) xylem vessels **(1)**
   (c) lignin **(1)**
3 *1 mark for the quality of the drawing, in a low power plan no cells should be drawn. Recognisable vascular bundle, xylem, cambium, phloem and sclerenchyma drawn for 3 marks, 3 or 4 drawn for 2 marks and 1 or 2 for 1 mark.*

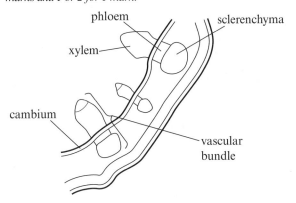

(4)

65. Looking at plant fibres

1 (a) This is valid for 5 mm coir fibres as strength of brown is likely to be no less than 343 − 36 MPa, that is, 307 MPa, whereas white is likely to be no more than 192 + 37 or 229 MPa at the 95% level of confidence. However, it is not valid for 35 mm coir fibres because the SDs overlap. It is also not valid because 5 mm white fibres are stronger than 35 mm brown fibres. **(3)**
   (b) Brown and white fibres from the same coconut would be taken. The fibre would be suspended between two fixed points and then masses added until the fibre broke. This would be repeated on five fibres of each type. The temperature of the room in which the experiments were conducted would be monitored and controlled as closely as possible. The masses would be added in the same way to each type and length of fibre. **(4)**
2 Bioplastic is made from starch from plants, which are themselves renewable. Oil-based plastic is from non-renewable resources. In addition, bioplastic is biodegradable and will not accumulate in the environment. **(3)**

66. Water and minerals

1 *This is designed to be a levels marked question in which credit is obtained for deployment of your knowledge and understanding of the topic. The bullet points below are called indicative content and as such you do not have to make all the points listed and you can make points which are not listed.*
   - Clear statement of dependent variable, that is, exactly what is to be measured: stated mass of plant tissue, mass of fruit, length of shoot, (number/colour) of leaves.
   - Clear description of method of measuring change in dependent variable.
   - Clear statement of independent variable (concentration of calcium).
   - Range of suitable concentrations suggested (at least five).
   - Some clear consideration of time period over which the growth will be measured.
   - Identification of other variables that could affect growth.
   - Description of how identified variables can be controlled.
   - Idea of need for replication at each concentration.
   - Control of source of plant, for example, use of same species/variety/source of seeds.
   - Use of graph to identify other values of concentration to test to identify optimum concentration.

For example:
I will be using tomato plants.
In this investigation, the dependent variable will be the number of leaves produced over a known time period. The number of leaves will be counted every week for a period of ten weeks. The tomato plants will be grown in pots containing soil-less compost to which 100 cm³ of a 1% calcium carbonate solution is added once a week. Distilled water is used to water the plants as and when needed. The plants will be grown in a controlled environment chamber in which the temperature is kept at 20 °C and light is provided by fluorescent tubes that are on for 12 hours per day. The tomato plants will be of the same variety and all grown from the same packet of seeds. The experiment will be repeated at concentrations of 2%, 3%, 4% and 5% calcium carbonate solution.
The results will be plotted on a graph with *x*-axis calcium carbonate concentration and *y*-axis number of leaves. If the graph does not level off, higher concentrations of calcium carbonate will be tested until the point at which the growth levels are found. This point shows the optimum concentration of calcium to be used. Beyond this concentration it would be uneconomic to add more calcium and achieve no better growth. **(6)**

2

Feature	Example of importance to living things
high specific heat capacity	keeps the temperature in water bodies fairly constant from season to season
polar solvent	allows the transport of nearly all biologically important substances.
high surface tension	a 'skin' over water that helps organisms move on its surface; it also helps in plant transport
incompressibility	used in hydraulic systems in living things like hydroskeletons in starfish
maximum density at 4 °C	allows aquatic organisms to carry on with their life even when temperatures are below freezing

(5)

67. Developing drugs from plants

1 Withering's study was less reliable because there was a smaller number of patients tested. Withering did not use a double-blind trial or compare treated patients with those given a placebo. **(3)**
2 The double blind trial involves some people with the new drug and some without it. The drug is replaced with a placebo, such as a sugar-coated dummy pill. Neither the doctors nor the subjects know who is on the new drug and who is not. This method helps to see if the new drug works better than the placebo and eliminates bias. **(3)**
3 (i) B
   (ii) B

68. Investigating antimicrobial properties of plants

1 All the extracts have an effect on all the bacteria as shown by the clear zones in every case. Apple was most effective against A whereas guava is most effective against B and C. Orange is more effective against A and C than pomegranate. Bacterial species A is least affected by guava bacterial, species B least a affected and bacterial species C is least a affected by apple and pomegranate. **(4)**
2 An extract is made from seeds of Jatobá. An agar plate with a culture of bacteria grown in nutrient broth is made. The broth has been previously sterilised in an autoclave. The extract of Jatobá seeds is placed on paper discs. A control is made with the solvent liquid but no extract added to discs. The plates are incubated at 25 °C for two days. After this time a zone of inhibition is looked for. **(5)**

69. Conservation: zoos

1 90% genetic diversity keeps many alleles in the population, thus providing a wide variety of phenotypes. The population needs to show a lot of phenotypic variation as the tiger lives in a wide range of habitats in the wild. Also, if there were a change in the environment the tigers would be unlikely to be able to adapt if genetic variety was low. **(4)**

2 Some people think that animals behave unnaturally in zoos. Also, a high proportion of animals kept in zoos are species that are not endangered. Some people feel that animals are kept in poor conditions in at least some zoos. **(3)**

3 The genetic diversity is reduced as there is a smaller gene pool. There will be fewer alleles due to inbreeding. **(4)**

70. Conservation: seed banks

1 As the world changes, not least due to a change in climate, existing crop strains may become less successful. In the case of wheat, there maybe increased drought stress or new pests strains may arise. Genes found in ancient strains, which have been eliminated from done ones, may prove useful in combatting threats to wheat yields.

2 Seed banks are advantageous because the seeds are stored in cool and dry conditions which means they can be stored for a long time. This is less costly than conserving living plants so large numbers of plants can be stored. Also, they take up less space and require less care. Their viability can be tested at regular intervals. The species is less likely to be damaged by natural disasters and disease. **(6)**

3 The wild plants may carry genes that can be used in crop plants to confer resistance to pests and diseases.
The plant may produce a chemical which might be useful in medicine, such as an anti-cancer agent. **(3)**

71. Exam skills

(a) (i) 1 elephant per 2 km^2 would be best, that is a population of 33 000 elephants

(ii) In 1980 the elephant population was greater than the 0.5 km^{-2} required to achieve maximum biodiversity, so it would be below maximum. From 1980 until 1985 the population declined – as the elephant population fell, so biodiversity would have risen. Just before 1985 however, the population fell below the optimum, leading to a fall in biodiversity again. From the mid 80s until the early 90s biodiversity would remain below maximum, although only slightly. By 1995, probably due to the new law protecting elephants, the population had exceeded the optimum level for biodiversity, so it would have started to decline again. It would be likely to reach a minimum for the two decades in 2000, when the elephant population was almost 3 times higher than the level giving maximum biodiversity.

(iii) Data on the number of different species would be needed. Various sampling methods including quadrats, traps, netting and various other techniques could achieve this. Also, an estimate of the numbers of individuals within each species would be needed. For this, quantitative methods would be needed such as counts in known areas (e.g. quadrats) or mark/recapture studies. The data could then be substituted into the formula :

$$D = \frac{N(N-1)}{\Sigma n(n-1)}$$

Where D = the index, N = total number of organisms of all species, n = number of organisms of each particular species.

72. Ecosystem ecology

1 All ecosystems must have a source of energy so this bottle would need to be placed in the light (assuming the producer is photosynthetic). The microbial culture probably has a photosynthetic alga as a producer. But the system also needs decomposers otherwise all the nutrients would become locked up in dead bodies. So the other type of organism in the microbial culture is probably a decomposer, such as a bacterial species. The fertiliser will provide nutrients, other than water and carbon dioxide, which are needed for algal growth. The brine shrimp is the consumer in this system. **(4)**

2 A population is made up of individuals all of the same species and interacting with each other. The most obvious interaction is between males and females during reproduction. A community consists of a number of populations in the whole ecosystem of animals, plants and microorganisms. Again, these populations interact with each other. Some examples of interactions are predator-prey relationships, pollination and mutualism. The communities within ecosystems are often referred to by their subsets. This may be by taxonomic grouping (for example, the plant community) or by habitat (for example, the soil community). **(4)**

3 Light is a factor and – where the area is covered by the canopy of the forest there will be a lower light intensity. Plants living in such regions will have reduced photosynthetic rate and thus may be less common in such areas.

73. Distribution and abundance

1 (a) A transect line would be laid out at a right angle to the front edge of the glacier. Quadrats would be placed at intervals along the transect line. These could be regularly sampled every 5 m. Abundance of *E. latifolium* would be estimated using a percentage cover method or by counting individual plants. Repeat transects could be carried out in order to measure the reliability of the data. All the percentage cover estimates would be recorded in a table next to the position along the transect line. **(4)**

(b) A light meter would be used. It would be held above the *E. latifolium* leaves but below that of any plants shading it. Readings would be taken along the transect line at each of the positions where *E. latifolium* was sampled. Several readings would be taken at each position to measure reliability. **(4)**

2 B **(1)**

74. Exam skills

1 The graph clearly shows a change in the tree species over time. At the beginning there are no trees present and it is not until 40 years that trees appear in the form of willow and alder. The changes that occur are brought about by the trees themselves. Each stage, alder and willow, then alder, willow and spruce, alder, willow, spruce and hemlock spruce, is referred to as a sere. Eventually, it is clear that the spruce, which arrives at 80 years, outcompetes the alder and willow, which have disappeared by 200 years, alder by 120. For at least the final 40 years, and possibly longer, the area has remained unchanged. It is dominated by spruce with a small amount of hemlock spruce. This is the climax community. **(4)**

2 (a) GPP = respiration + net primary production; this is given by all the grass passed on to other organisms, field mice, grasshoppers and seed-eating birds = 1980 + 3600 + 11.6 + 44.4 + 6, which is 5642 kJ m^{-2} year^{-1} × 10^4 **(3)**

(b) energy taken in = 44.4
energy lost = 12.5 + 25.4 = 37.9
so energy converted into new grasshopper = 44.4 − 37.9 = 6.5 = (6.5 ÷ 44.4) × 100 % = 14.6% **(3)**

75. Succession

1 First of all, there would be a large decrease in the number of plants because they would be covered by ash. After this, the area would undergo a process of succession in which the soil would improve. Sometime after the eruption, pioneer species, such as lichens, would grow on the cooled lava. Due to soil improvement, low-growing plants, such as ferns and grasses, would be able to grow. These would shade out the lichens and other lower-growing species. This would lead to an increase in the number of plant species. **(4)**

2 A climax community is the final sere of the successional process. It is self-sustaining and stable, usually with one dominant species. **(3)**

76. Productivity in an ecosystem

1 phytoplankton is 52% of total
so is $840 \times 10^6 \times 0.52 \, \text{kJ m}^{-2} \, \text{year}^{-1}$
$= 437 \times 10^6 \, \text{kJ m}^{-2} \, \text{year}^{-1}$ **(2)**

2 The higher the temperature the more NPP, and NPP depends on photosynthesis. There are enzymes involved in photosynthesis. The enzyme-catalysed reaction has more kinetic energy and can work faster at higher temperatures. An increase in rainfall increases NPP. This might be because water is needed for light-dependent reactions as well as the transport of mineral ions, amino acids and sucrose. **(5)**

77. Energy flow

1 Sun to producers: $(7600 \div 200\,000) \times 100 = 3.8\%$
producers to primary consumers: $(750 \div 7600) \times 100 = 9.9\%$
The efficiency between producers and primary consumers is $9.9 \div 3.8 = 2.6$ times greater than that between the Sun and the producers. **(4)**

2 plants to primary producers: $(240 \div 2250) \times 100 = 10.7\%$
primary producers to secondary producers: $(38 \div 240) \times 100 = 15.8\%$ **(2)**

78. Photosynthesis – an overview

1

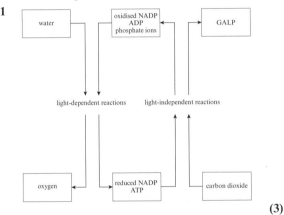

(3)

2 A **(1)**

3 C **(1)**

79. The light-dependent reactions

1 Energy from light splits water in a process called photolysis. This gives hydroxyl ions, which split up to give protons and oxygen. **(3)**

2 The electrons are passed into a chlorophyll molecule to fill 'electron holes' in it. Light energy then excites the electrons and promotes them to a higher energy level. From an electron acceptor they are passed down a chain of carriers; the energy they release being used in ATP production. They are finally accepted by NADP to make NADPH. **(4)**

3 (a) A **(1)**
   (b) D **(1)**

80. The light-independent reactions

1 There will be a reduced production of carbohydrate because less carbon dioxide will be fixed in combination with RuBP. This is because the joining of carbon dioxide and RuBP is catalysed by RuBP carboxylase. The Calvin cycle will run slowly because the first step, carbon fixation, is inhibited, giving less carbon for reduction to carbohydrate. **(4)**

2 (a) RuBP has 5 C atoms, so 6 would have thirty. GP has 3 C so 12 have 36. The difference is therefore 6 C atoms. **(2)**
   (b) This is because 2 molecules of the GP are used to make 1 molecule of glucose, which has 6 C atoms. **(2)**

3 B **(1)**

81. Exam skills

1 (a)
   • Succession is the sequence of species and communities replacing each other over time.
   • Quadrat 1 is in the youngest part of the system and 9 in the oldest.
   • There are different communities of plants at different distances.
   • There are only one or two near the beach.
   • Sea couch would appear to be a pioneer species.
   • Organic matter content increases with distance from the beach.
   • This is likely to lead to increased water-holding capacity and mineral content.
   • There are more species further from beach, with 18 being the most.
   • Competition from one species may lead to loss of another as time goes by, for example, there is no sea couch after quadrat 1 although it might be able to grow in the soil in quadrats 2 and beyond.
   • There are two species in the final sample.
   • The final stage might be climax community. **(9)**

   (b) A soil sample could be taken from each quadrat and a plant found in only a few quadrats, for example, sea couch, could be grown in this soil in controlled conditions. If it grows well then the soil is OK and this suggest its absence in nature is due to competition. **(3)**

82. Chloroplast

1 D **(1)**

2 (a) measured length = 66 mm
   size of real object = length ÷ 7500
   = 0.0088 mm = 8.8 μm
   1 mark for correct units for given answer: μm

   (b) R is the stroma is where the light-independent reactions occur.
   S is the granum is where the light-dependent reactions occur.
   T is the envelope, which controls the movement of substances in and out of the chloroplast. **(3)**

83. Climate change

1 The climate has become warmer shown by the fact that, between 8700 and 6390 years ago, larch and spruce were growing but then died out. Larch and spruce are only found in boreal and northern temperate regions in the present day. Boreal and northern temperate regions are relatively cold climates. Pine was not growing but has become established more recently. Pine is only found in southern boreal and temperate regions (in present day). Southern boreal and temperate regions are warmer climates. **(5)**

2 Carbon dioxide and methane are greenhouse gases. They both trap heat reflected from the Earth's surface, which increases its mean temperature. Carbon dioxide is relatively more abundant than methane but methane is the more powerful greenhouse gas. **(3)**

84. Anthropogenic climate change

1 Greenhouse gases in the atmosphere trap heat energy reflected from the Earth's surface.
Increased levels of these gases from fossil fuel combustion increase the greenhouse effect leading to an increase in mean temperature of the Earth's surface.
However, carbon neutral fuels, such as those derived from biomass, do not, so the statement is not completely true. **(4)**

2 (a)

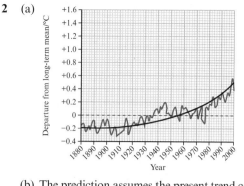

(1)

(b) The prediction assumes the present trend continues and this is not valid as the line is not based on a long enough series of data. In addition, the graph shows that temperature has fluctuated over the period 1880 to 2000. For example, there were big falls between 1900 and 1908, and again between 1945 and 1958. An unpredictable event, such as a volcanic eruption, may affect the trend. There may be a sharper than expected change in fossil fuel consumption. **(4)**

85. The impact of climate change on species

1　Rate is given by the gradient of each line
　　rate for 10 °C is $0.27 \div 20\,\text{cm}^3\text{min}^{-1} = 0.014\,\text{cm}^3\text{min}^{-1}$
　　rate for 20 °C is $0.49 \div 20\,\text{cm}^3\text{min}^{-1} = 0.025\,\text{cm}^3\text{min}^{-1}$
　　$Q_{10} = 0.025 \div 0.014 = 1.79$ **(3)**

2　The graph on the left shows that the temperature increased by 2 °C between 1950 and 2010. The graph on the right shows that the upper limit of both species has fallen. In the case of barnacles, the upper limit of intertidal height is now 280 cm as opposed to 310 cm, and for mussels it has gone down from 250 cm to 220 cm. There is a correlation between the rise in temperature and the fall in the upper limit of the two species. There is no evidence from this study, however, that one is the cause of the other. **(4)**

86. The effect of temperature on living things

1　Warming may lead to reduced hatching and so a reduced population of adults if the temperature exceeds 28 °C. Where the present temperature is currently less than 28 °C, global warming would increase hatching rate and populations. Earlier hatching may mean the population is out of synchronisation with food supply, causing a decrease. There could be an increased population due to reduced predation or a decreased population due to reduced food supply. **(3)**

2　(a) Oak usually opens before ash; ash only opens before oak when it is less than 6 °C. The warmer the spring, the bigger the gap between oak and ash because oak gets earlier and earlier, but ash stays more or less the same. There are only 3–6 days difference in a cold spring but over 35 days difference in a warm spring. **(2)**

　　(b) They would probably become less common because it would be too shady too soon. **(2)**

87. Decisions on climate change

1　A biofuel is any source of energy produced by recent photosynthesis. It provides a renewable energy source and is carbon dioxide neutral. **(2)**

2　When a forest is young, trees will be absorbing more carbon dioxide in photosynthesis than they are releasing in respiration, so they are net absorbers of carbon dioxide. In a mature forest, photosynthetic gain will equal respiratory loss and thus the forest will not be a net absorber, but it does store carbon fixed in its earlier years. **(4)**

3　(a) double the 2004 level of production is 60 million tonnes
　　85% of 60 is $(60 \div 100) \times 85$
　　= 51 million tonnes **(3)**

　　(b) Biofuel is less directly polluting and palms act as a carbon sink. In addition, it is a renewable resource conferring

economic benefits on the country. Palms provide new habitats, although they are avoided by orangutans, which are in danger of extinction. Although the palms form a forest, it has less biodiversity and less forest resources than what it replaces. The original forest may be a bigger carbon sink. **(4)**

88. Evolution by natural selection

1　A **(1)**
2　B **(1)**
3　D **(1)**
4　Nucleic acids are the genetic material in all organisms. **(1)**

89. Speciation

1　Sympatric speciation is when the populations are isolated in a way other than by geographical separation, and this leads to speciation. In this case, reproductive isolation has occurred due to the two distinctly different life cycle timings. Consequently, the two populations of flies in the wild do not have mature reproductive adults existing at the same time. **(9)**

2　A gene that was involved in the control of lip shape mutated. Also, one coding for height mutated as well. The change in lip shape resulted in better adaptation for feeding. Also, the greater height of white rhinos protected them in the open. The alleles for height and lip shape were passed on to offspring. This led to a change in allele frequency. The different food sources resulted in selection pressures. The rhinos would have become reproductively isolated. **(5)**

90. Death, decay and decomposition

1　(a) Temperature of the body core changes with time after death. The core temperature depends upon the ambient temperature. Other post-death changes, such as degree of muscle contraction and decomposition, also depend on ambient and body temperature. **(3)**

　　(b) For the clothed body the estimate was too short because the clothing would reduce heat loss due to trapping the heat. For a curled up body, the estimate was too short because heat loss is reduced as the exposed surface area was smaller. For a body in moving air, the estimate was too long because moving air speeds up heat loss. **(5)**

2　The estimate may be incorrect because there is no record of environmental conditions. The temperature may have varied, leading to variation in the development rate of insects. It may also affect the time taken to lay eggs. There is also no knowledge of the body's previous treatment. **(3)**

91. DNA profiling

1　(a) Substance X is (DNA)primer(s).
　　Substance Y is (mono)nucleotide(s).
　　Substance Z is DNA strand(s). **(3)**

　　(b) heated to 90–95 °C　　　T1
　　heated to 75 °C　　　　　T3
　　cooled to 55–60 °C　　　T2
　　cooled to 4 °C　　　　　none of them involve this stage **(3)**

2　The DNA is cut up using restriction enzymes and a sample placed in a well in the gel. A potential difference is applied across the gel to separate the fragments and a stain applied to visualise them. They show up as bands and the number of bands and position is unique to the organism from which the DNA came. **(4)**

3　comparing total number, position and width of bands **(3)**

92. Exam skills

1　(a) the proportion that one allele is of the total alleles of one gene in a population **(2)**

　　(b) The two different alleles must both exist at both sites (which can be assumed to have different copper ion levels in the soil). At the site with more copper ions, possession of the allele will confer an advantage. The selection pressure exerted by the presence of copper ions will mean individuals that possess the allele will be more likely to reproduce and pass the allele on to their offspring. **(4)**

   (c) Bacteria have a faster life cycle and are subject to greater selection pressure, such as antibiotic use. The larger numbers of bacteria means there is a larger gene pool. **(3)**

93. Bacteria and viruses

1 The genetic material of viruses is in the form of DNA or RNA. This is single-stranded (RNA) or double-stranded (DNA). The whole is surrounded by a capsid. There may be an envelope. They have glycoproteins on the outside and some viruses contain enzymes. **(4)**

2 *Any three points from:* Bacteria have these, but viruses do not:
- cell wall
- cell membrane
- ribosomes
- cytoplasm.
Bacteria are larger. **(3)**

3 (a) $800 + 600 = 1400$
$(800 \div 1400) \times 100 = 57\%$ **(2)**

   (b) A greater proportion had HIV in 2008 by $57.14 - 23.81 = 33.33\%$. **(2)**

94. Pathogen entry and non-specific immunity

1 The gut flora prevents growth of, or kills, harmful bacteria by competing with them for space and nutrients. They are also thought to release toxins. **(3)**

2 The immediate effect is to remove all bacteria except G within seven days. This suggests G is resistant to the antibiotics. All the other types, except G, are sensitive to the antibiotics. After this, G is the only type of bacterium from seven days until nine months. Then there is an increase to four different types of bacteria at 12 months. Three are ones that were there at the start and one is a new type of bacterium, J. This new type has probably come from the diet. At 18 months, there are the same four types of bacteria. G has the highest percentage at 18 months, compared to H, I and J, possibly because G can outcompete the others for space and nutrients. **(5)**

95. Specific immune response – humoral

1

Description	True	False
B and T cells are formed in the bone marrow	✓	
B cells stimulate T cells to produce clones of memory cells		✓
T helper cells produce chemicals that destroy pathogens		✓
B and T cells are able to form clones by mitosis	✓	

(4)

2 This is an example of artificial active immunity. A vaccine could be made containing the synthetic molecule, which acts as an antigen. Its presence will stimulate the specific immune response to the synthetic antigen eventually leading to clonal expansion of B cells, which will produce 2G12 antibodies. Memory cells will also be made which will lead to faster production of 2G12 antibodies on reinfection. **(4)**

3 The bacterium is made of many different chemicals that can act as antigens.
B lymphocytes recognise specific antigens and will be activated by T lymphocytes. B lymphocytes undergo mitosis to form genetically-identical plasma cells which produce the antibodies. **(4)**

96. Specific immune response – cell-mediated

1 The immune response will be weaker and the person may not recover from this infection. Cytokines are released from T helper cells and are involved in activation of B cells and T killer cells. Impaired B cell function will lead to the production of fewer antibodies. **(4)**

2 (a) host cell infected with virus **(1)**
   (b) cytokine **(1)**
   (c) mitosis **(1)**
   (d) T killer cells are involved in the destruction of virus-infected host cells by perforins. Adenovirus particles are released from cells and antibodies can now bind to them. The virus can now be destroyed by macrophages, and memory T killer cells are produced ready for secondary immune response. **(4)**

97. Post-transcriptional changes to mRNA

1 13 exons and 12 introns **(1)**

2 Post-transcription modification of pre mRNA can give rise to a number of different mRNA molecules, which would be translated into different proteins.
Mutation gives rise to a different gene, which in turn gives different mRNA and thus a different protein. **(4)**

3 Enzymes are made of protein, so the theory really says one gene-one protein. Since the information says that some genes code for over 1000 proteins and the average is five, this theory is shown to be incorrect. A better idea would be the one gene-one family of related proteins. **(3)**

98. Types of immunity

1

Immunity	Active	Passive
natural	Q	P
artificial	R	S

(4)

2 (a) This shows the secondary immune response. This suggests antibody A was in the vaccine and caused a primary response that led to the production of memory cells. On infection, the memory cells are activated, leading to rapid production of plasma cells and therefore antibodies. **(3)**

   (b) Antibodies will only be made if the antigen is present. Therefore, antigen B was not present in the vaccine, which is why vaccination failed to stimulate immune response. **(2)**

99. Antibiotics

1 (a) because they do not have cell walls **(1)**
   (b) Tetracycline acts on ribosomes, which are involved in the manufacture of protein. Proteins are vital as enzymes in the membrane, amongst other things, so lack of them will cause cells to die or be unable to reproduce. **(2)**

2 (a) Cephalosporin kills bacteria because the cells burst. **(2)**
   (b) The cells cannot replicate as there can be no cell division if DNA cannot be replicated. **(2)**

3 (a) wire loop/sterile pipette **(1)**
   (b) Agar powder is dissolved in distilled water. The agar solution is autoclaved. On removal from the autoclave it is cooled until it sets, but then re-melted before pouring into a Petri dish. The dish is swirled to ensure even coverage. An inoculating loop is flamed to sterilise it; this is an aspect of aseptic technique. It can then be used to pick up some of the broth, which can then be streaked on to the agar. The lid is kept close to the plate during streaking. The loop is sterilised and cooled between each set of streaks. **(4)**

100. Evolutionary race

1 (a) The fall from 2000 to 2006 is $226 - 142 = 84$, which is $(84 \div 226) \times 100 = 37\%$. So, in six years, the average annual percentage fall is $37 \div 6 = 6.2\%$. If this were to go on a further six years until 2012, it would fall a further $6 \times 6.2\% = 37.2\%$. This, plus the 37 already achieved, gives 74.2, well above the 43.3% target. **(5)**

(b) This is because some bacteria are resistant to antibiotics and this resistance can be passed on to future generations. MRSA is a well-known example. **(2)**

2 A mutation has occurred in the DNA that could lead to a change in the outer surface, and therefore antigen, of the bacteria. Memory T cells will not recognise the new antigen. So another primary immune response is needed to activate another population of T helper cells. Phagocytes may be unable to recognise and engulf the *Mycobacterium tuberculosis* so antigen presentation is not possible. **(4)**

101. Exam skills

1 (a) There is a clear correlation in the graphs between increasing use and increasing resistance. The increased use of the antibiotic means that more bacteria are exposed to it. Some of these bacteria may have undergone mutation that gives them an advantage over others due to resistance to the antibiotic. The ones with the resistance mutation are selected for, replicate and pass the resistance allele on to their offspring. These antibiotic-resistant organisms become more common. The resistant allele may also be passed to other bacteria by the plasmids or conjugation, a form of sexual reproduction in bacteria. **(6)**

(b) Although the clear area around antibiotic B is larger than that around antibiotic C, this may not be due to its greater effectiveness. For example, if antibiotic B is a smaller molecule than antibiotic C, the larger clear area may be simply due to it having diffused further through the agar in the time. **(3)**

102. Skeleton and joints

1 Because muscles can only generate movement by pulling. All joints that allow movement are made in such a way as to allow movement one way when one of the muscle pair contracts and the other way when the other muscle contracts. **(3)**

2 (a) D **(1)**
   (b) B **(1)**

3 When the mass is steady, muscles A and B are contracting. Joint D will be at right angles. On lifting, B will contract further and A will relax. D will become more acutely angled on the inside. **(3)**

103. Muscles

1 (a) C **(1)**
   (b) B **(1)**
   (c) G **(1)**
   (d) A **(1)**
   (e) A **(1)**

2 ATP binds to the myosin head and, as a result, the myosin head detaches from the actin. The ATP is hydrolysed to form ADP and Pi and this hydrolysis causes a change in the shape of the myosin head, returning it to its upright position. **(3)**

104. The main stages of aerobic respiration

1 (a) 1 TP requires 1 ATP to be made so this is −1
      1 TP yields 2ATP so −1 + 2 = 1
      Krebs yields 1 ATP from each TP so this is 1 + 1 = 2
      1 TP yields 5 NADH, each of these yields 3 ATP, so this is 15 ATP
      1 TP yields 1 $FADH_2$, which yields 2 ATP
      so 1 TP yields 2 + 15 + 2 = 19 ATP
      × 2 for 2 TP = 38 ATP **(3)**

   (b) $FADH_2$ and NADH pass H^+ to electron transport chain carriers. In the inner mitochondrial membrane, H^+ is passed into the inter membrane space and then passes back through ATPase converting ADP + Pi to ATP in oxidative phosphorylation by chemiosmosis. **(4)**

105. Glycolysis

1 (a) R is ATP
      S is ADP
      Q is ATP
      V is ADP
      X is NADH **(5)**

   (b) R is 2 molecules; Q is 4 molecules **(2)**

   (c) This enables glucose to react. It does this by lowering activation energy and also maintains a concentration gradient preventing loss of glucose from the cell. **(3)**

106. Link reaction and Krebs cycle

1 (a) A is pyruvate; B is glycolysis **(2)**

   (b) carbon dioxide, as the compounds it is made from have one fewer carbon each time **(2)**

   (c) hydrogen **(1)**

   (d) The cycle would stop and the 4-carbon compound would accumulate. At the same time the 6-carbon compound would run short as would the 5-carbon compound. All this leads to a reduction in the production of molecules T and H. **(3)**

107. Oxidative phosphorylation

1 (a) ATPase **(1)**

   (b) The gradient is from the inter membrane space into the matrix, so needs a high concentration in the space. H^+ ions from reduced NAD are pumped into the inter membrane space, requiring energy derived from movement of electrons along ETC on the cristae. **(4)**

   (c) Electrons are passed along the electron transport chain during which they lose energy. This adds a phosphate to ADP to make ATP via ATPase. The final acceptor of these electrons is oxygen. **(3)**

108. Anaerobic respiration

1 Anaerobic respiration involves the breakdown of glucose to pyruvate in glycolysis. First the glucose is phosphorylated using two molecules of ATP. It can then be split into two molecules of phosphorylated 3-carbon compounds. Each of these is then oxidised, passing its electrons to NAD to give NADH. Phosphate from the phosphorylated 3-carbon compounds is used to phosphorylate ADP to form ATP. This happens twice for each, giving four ATP in total. Since two ATP were used at the start, this gives a net yield of two ATP per glucose molecule. In the absence of oxygen pyruvate, two electrons from NAD are used to make pyruvate into lactate, leading to the regeneration of NAD, which is necessary for glycolysis to continue. **(5)**

2 Most lactate is converted back into pyruvate. It is oxidised directly to carbon dioxide and water via the Krebs cycle, thus releasing energy to synthesise ATP. This means that oxygen uptake is greater than normal. This is called the oxygen debt and is needed to fuel the oxidation of lactate. **(3)**

3 The runner will need $60 \times 0.3\,dm^3$ to run for 1 minute. This is $18\,dm^3$. In that minute he can take in $3.5\,dm^3\,min^{-1}$ and so will need an extra $18 - 3.5\,dm^3 = 14.5\,dm^3\,min^{-1}$, which is the entire debt. He will have run $60 \times 5\,m\,s^{-1} = 300\,m$ in that time. **(3)**

109. The rate of respiration

1 (a) The carbon dioxide produced in respiration is absorbed. Otherwise, it would affect the volume and thus the pressure of gas in the equipment. This allows the measurement of oxygen used. **(2)**

   (b) The tap allows the flow of air to be changed from side arm to chamber to syringe to chamber. The syringe can then be used to push the coloured liquid back to the start. **(2)**

   (c) The mean distance for insects is $48\,mm \div 4 = 12\,mm$. So the mean rate is $12 \div 15 = 0.8\,mm\,min^{-1}$.

   (d) There is no oxygen available so there must be anaerobic respiration, which produced carbon dioxide. This is absorbed by the soda lime so there is no net change of volume or pressure of gas. **(2)**

110. Control of the heartbeat

1 (a) Electrical activity travels through the atria and reaches the AVN. The atria contract whereby their volume decreases and pressure rises forcing blood into the ventricles. **(3)**

(b) $60 \div 0.75$
$= 80$ **(2)**

2 A is SAN; B is AVN; C is Purkyne fibres **(3)**

111. Cardiac output and ventilation rate

1 SAN activity increases and thus the heart rate increases. More blood returns to the heart and causes heart muscle to stretch, increasing stroke volume due to ventricles contracting with greater force. The AVN time delay decreases. **(4)**

2 (a) age and gender **(2)**
(other acceptable factors are: level of fitness, resting pulse rate, diet, weight, heart conditions, blood pressure, drugs, height)

(b) output before training $= 152 \times 85 = 12\,920\,cm^3$
output after training $= 142 \times 96 = 13\,632\,cm^3$
increase $= 13\,632 - 12\,920 = 712\,cm^3$
$= 712 \div 1000\,dm^3$
$= 0.71\,dm^3\,min^{-1}$ **(3)**

(c) cardiac output is heart rate × stroke volume
The heart takes 0.7 seconds to refill so the rate is
$60 \div 0.7 = 85.7$.
The stroke volume is $148 - 55 = 93\,cm^3$
so output $= 93 \times 85.7 = 7970\,cm^3$ **(2)**

112. Spirometry

1 There are six peaks and troughs in 30 seconds so the breathing rate is 12/min.
The distance between a trough and a peak is the tidal volume and is $0.45\,dm^3$. **(2)**

2 (a) minute volume is tidal volume × number of breaths in a minute
at rest minute volume $= 0.4 \times 12 = 4.8\,dm^3\,min^{-1}$
during exercise $= 1.2 \times 36 = 43.2\,dm^3\,min^{-1}$
therefore, increase $= 43.2 - 4.8\,dm^3\,min^{-1} = 38.4\,dm^3\,min^{-1}$ **(3)**

(b) Oxygen consumption is given by the slope of the curve. At 6 seconds the peak is $3.5\,dm^3$ and at 35 seconds it is $1.4\,dm^3$. So $3.5 - 1.4\,dm^3$ of oxygen is consumed in 29 seconds which is $2.1\,dm^3$ in 29 seconds. This is $2.1 \div 29\,dm^3\,sec^{-1} = 0.072\,dm^3\,sec^{-1}$. Thus per minute rate is $0.072 \times 60 = 4.34\,dm^3\,min^{-1}$. **(3)**

113. Fast and slow twitch muscle

1 (a) The change in sensitivity for fast twitch muscle between pH 7 and pH 6 is 0.6 a.u. For slow twitch, the reading at pH 7 is 1.0, and the reading at pH 6 is 2.0, so change in sensitivity is 1.0 a.u. Both are less sensitive to calcium at a lower pH but at the higher pH the effect on slow twitch is greater. In both, a lower pH decreases contraction and in both a lower pH has no effect at high calcium ion concentration. **(4)**

(b) Fast twitch is anaerobic, slow twitch is aerobic. In the body, fast twitch is more likely to experience low pH due to lactate from anaerobiosis. It is less affected by change in pH so can continue to respond to stimulus at lower pH. **(3)**

2 They have less myoglobin and less haemoglobin as there are fewer capillaries present because respiration is mainly anaerobic. **(2)**

114. Homeostasis

1 A deviation from the normal level triggers a mechanism to eliminate the deviation. An increased level of cortisol inhibits CRH secretion from the hypothalamus. This reduces secretion of ACTH from the (anterior) pituitary which causes a drop in the level of cortisol. **(3)**

2 An increase or decrease of carbon dioxide in plasma causes a decrease or increase in pH due to the formation of carbonic acid. These changes are detected by chemoreceptors in the medulla and carotid and aortic bodies. From these, impulses are sent to the respiratory centre, which in turn sends impulses to the intercostal muscles and diaphragm, increasing or decreasing breathing rate and depth. **(4)**

115. Thermoregulation

1 The fall in temperature is detected by receptors in the hypothalamus. There will be increased muscular activity in the form of shivering, reduced sweating, and increased metabolic rate and behavioural responses such as wearing a sweater. The metabolic rate will increase and less blood will flow to the extremities due to peripheral vasoconstriction leading to a reduction in heat loss by radiation. Brown fat may also be hydrolysed. **(4)**

2 Heat energy from blood in the capillaries is absorbed by sweat and used to break hydrogen bonds in water, creating water vapour. The energy needed to turn water into water vapour is called latent heat of vaporisation. This takes heat from the body. **(3)**

3 A change in body temperature is detected by the hypothalamus via its thermoreceptors.
As a consequence, there are changes in the production or loss of heat such as sweating rate and metabolism. A return to normal temperature switches the regulatory mechanism off. So a rise in temperature will cause a fall in temperature which will turn off mechanisms reducing temperature; this is negative feedback. If a rise caused a further rise this would be positive feedback. **(5)**

4 D, the hypothalamus **(2)**

116. Exercise

1 The benefits include: a lowering of blood pressure due to arterial vasodilation of exercise, an improvement of the ratio of HDL to LDL, a reduction in the likelihood or extent of obesity, and prevention of type II diabetes.
Potential dangers are to the immune system, which is suppressed by both too little or too much exercise. Moderate exercise appears to stimulate the immune system. In those who exercise too much there are fewer natural killer cells. Also, inflammatory responses in muscles often reduce non-specific immune responses elsewhere.
Exercising too much is also associated with joint damage, particularly of the cruciate ligament in the knee. **(6)**

2 Obesity is linked to increased risk of diseases, such as diabetes and CVD. Treating such disease will put an economic burden on society. **(3)**

3 With an efficient metabolism, less food is required to deliver the energy requirement. This means that such a person would be more likely to have extra food not respired, which could be laid down in the body as fat. **(2)**

117. Sports participation and doping

1 (a) (i) **D**
(ii) **B (2)**

(b) With keyhole surgery there is a small incision, so there is less bleeding, less damage to tissue, reduced chance of infection and a shorter recovery time. For society, it means that more patients can be treated in a given time and is cheaper than invasive surgery, and thus has economic advantages. In addition, costly anaesthesia can be avoided using this technique. **(3)**

2 (a) 0.96% **(1)**

(b) Drugs like amphetamines can cause long-term adverse effects to the user. In addition, some would say this would confer unfair advantage on the users. On the other hand many would maintain that individuals have the right to make their own choices and a responsibility to live with the consequences, so a ban is unnecessary. **(3)**

118. Exam skills

1 (a) Enzyme 1 converts DHAP to 2-PG. 2-PG then becomes the substrate for the next reaction where another enzyme, enzyme 2, converts it to PEP. There will be only one enzyme which will allow 2-PG to fit into its active site, that is, it is specific. This enzyme speeds up and controls the conversion by lowering the activation energy. A third enzyme, 3, now uses PEP as a substrate and converts it into Pyruvate. Pyruvate will only result if all the enzymes are active. **(5)**

(b) W is NADox and is formed when reduced NAD releases electrons, which pass to carrier A. At the same time, H^+ are moved into inter membrane space. **(3)**

(c) Substance X is ATP and it is made when hydrogen ions pass through the stalked particle, ATP synthase. They do so via an electrochemical gradient releasing enough energy to join ADP to inorganic phosphate in a process called chemiosmosis. **(3)**

119. Mammalian nervous system

1

Feature	Type of neurone		
	Sensory	Relay	Motor
myelinated	✓	✓	✓
central cell body	✓	✗	✗
terminal cell body	✗	✓	✓
partially surrounded by Schwann cells	✓	✓	✓
found only in CNS	✗	✓	✓

(3)

2 (a) Y **(1)**

(b) A is motor; B and C are relay; D is sensory **(3)**

3

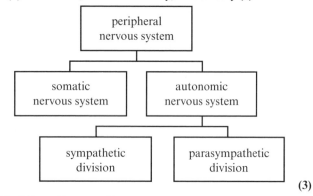

(3)

120. Stimulus and response

1 (a) pigment at the back of the eye absorbs light so none is reflected out **(1)**

(b) Depolarisation of the bipolar neurones is passed on via synapses to sensory neurones. These join together to form the optic nerve, which again will pass impulses on to motor neurones connected to the radial muscles of the iris. These will contract and cause an increase in pupil diameter. **(3)**

2 (a) As light intensity increases there is initially little change then pupil size decreases and then levels off. The line is sigmoid. **(2)**

(b) receptor: any photoreceptor, rods or cones
effector: either the radial or circular muscles of the iris **(2)**

121. The resting potential

1 (a) −70mV **(1)**

(b) by using microelectrodes, one inside the axon and one outside **(2)**

(c) The resting potential results from two opposing forces on K^+ ions. There is a large diffusion gradient for K^+ to leave the cell taking a positive charge. The inside is left more negative than outside. Eventually, the negative charge in the cell stops diffusion of K^+. The two opposing forces balance at −70mV. **(4)**

(d) The sodium ion channels close; sodium ions pumped out; use of ATP; potassium ions diffuse out; along concentration/electrochemical gradient; inside becomes negative relative to the outside. **(3)**

122. Action potential

1 (a) 0.90ms **(1)**

(b) 3.0ms **(1)**

(c) From the graph, it is 3.6ms before the cell is back to resting potential so another action potential could be initiated. So there could be 1000 ÷ 3.6 action potentials per second. This works out at 278. **(4)**

2 (a) A is sodium; D is potassium **(2)**

(b) At this point there is a diffusion gradient of potassium ions as well as an electrochemical gradient. The ions can move because there is increased permeability of the membrane to potassium ions. The potassium gates are open whereas the sodium gates are closed. **(3)**

123. Propagation of an action potential

1 Myelination allows impulses to travel faster due to its properties as an electrical insulator. The myelin is composed of Schwann cells wrapped around the axon. It is not continuous; there are regular gaps called nodes of Ranvier. Impulses jump from one of these nodes to the next, a process known as saltatory conduction. **(4)**

2 Local currents are set up and there is depolarisation in the next section of the axon. Sodium channels open and sodium ions move into the axon. As more sodium ions move in, more sodium channels open in an example of positive feedback. The axon becomes +40mV inside. The potassium channels open and potassium ions move out of the axon, causing the membrane to repolarise. The wave of depolarisation-repolarisation repeats along the axon. The impulse moves in one direction only and the current jumps between nodes of Ranvier gaps in the myelin sheath, speeding up conduction by the saltatory effect. **(5)**

3 *Choose 2 to 4 and 200 to 400.*
In the myelinated axon, it goes from 20 to 42 so increase is 22 so (22 ÷ 20) × 100 = 110%.
In the unmyelinated axon, it goes from 12 to 20 so increase is 8 so (8 ÷ 12) × 100 = 67%. **(3)**

124. Synapses

1 Bungarotoxin may prevent the release of neurotransmitter from the pre-synaptic membrane. Alternatively, it may have a shape similar to that of the neurotransmitter. In this case, it would bind to and consequently block the receptor for the neurotransmitter on the post-synaptic membrane. This may lead to permanent depolarisation of the post-synaptic membrane and an inhibition of action potentials. Another possibility is that it inhibits the breakdown of the neurotransmitter and is itself unaffected by the enzyme that does this job. **(4)**

2 The neurotransmitter diffuses across the synaptic cleft and binds to receptors on the postsynaptic membrane. Gated channels open, allowing sodium ions to pass through the postsynaptic membrane, causing depolarisation. If sufficient depolarisation occurs an action potential is set up in the adjacent cell. All of this allows a one-way flow of information from one nerve cell to the next. The neurotransmitter is now broken down so that the prolonged action potential in the post-synaptic membrane does not occur. The products of this breakdown would be reabsorbed through the pre-synaptic membrane. **(6)**

3 It is most likely that atropine has a similar shape to acetylcholine. This means it would bind to receptors, thus preventing acetylcholine from binding and stopping depolarisation of the post-synaptic membrane. **(2)**

125. Vision

1 (a) **D (1)**
 (b) **A (1)**
 (c) **C (1)**
2 rhodopsin + light → opsin + retinal **(2)**
3 Rhodopsin consists of retinal and opsin. Light energy is absorbed by rhodopsin, which causes retinal to change shape from the cis to the trans form. As a result, rhodopsin splits and is said to be bleached. Rod cell sodium gates close so less sodium diffuses in, resulting in hyperpolarisation. The bipolar cell becomes depolarised, forming an action potential in the ganglion cell. **(4)**

126. Plant responses

1 (a) leaves or any plant part **(1)**
 (b) There is a slow conversion in darkness from P_{FR} to PR. Far-red light causes a fast conversion of P_{FR} to PR. **(2)**
 (c) exposure to red light wavelength 660–700 nm **(1)**
2 Both sides of the shoot would be taller than a control. In both cases, IAA diffuses down to the zone of the elongation and stimulates cell elongation. This is brought about by an increase in the permeability of the cell wall to water which enters, swelling the cell. The shoot would bend to the right due to the increased growth on the left-hand side caused by the artificial IAA. **(6)**

127. Nervous and hormonal control

1 In hormonal control effect is longer lasting and slower. It is also often irreversible. It involves chemical processes whereas nervous involves electrical. **(4)**
2 It binds to receptor on the cell surface directly. A second messenger molecule is then produced and this might affect gene expression. **(3)**
3 **B (1)**
4 **C (1)**

128. The human brain

1 (a) The images give information about the location, size and accessibility to surgery of the abnormality. **(3)**
 (b) fMRI detects levels of oxygenation of the blood and can therefore measure changes in blood flow within the brain. An increased flow would suggest an increase in activity. Such changes in activity can be followed whilst the subject carries out tasks or experiences stimuli. **(3)**
 (c)
regulating core temperature	X
climbing stairs	Z
regulating carbon dioxide in the blood	Y
choosing a gift	W **(4)**

129. Critical window for development

1 (a) (i) The kitten's visual system is similar to humans so the results can be extrapolated to humans without actually using humans, which would create greater ethical problems. **(2)**
  (ii) This reduces any genetic variation and therefore error in the experiment. **(2)**
 (b) (i) Rhodopsin splits into opsin and trans retinal. This changes the shape of retinal. **(2)**
  (ii) The lack of stimulus in the right eye would mean that fewer impulses would reach the visual cortex. This would lead to less neurotransmitter release and a weakening of the synapses. The neurones for the right eye would be lost. The result would be that the ocular columns would be smaller for the right eye than for the left eye. **(3)**
 (c) These animals feel pain. **(1)**

130. Learning

1 The mean degree of muscle contraction decreases. Calcium ion channels are less responsive in the sensory neurone so fewer Ca^{2+} are taken up and less neurotransmitter is released. This will result in fewer impulses passing along the motor neurone to the muscle. **(4)**
2 (a)

(4)

 (b) The critical value is 0.79 and the calculated value of 0.93 is greater than this at the 95% confidence level. This shows that there is a significant negative correlation between the number of taps received by the snail and the time it takes for the snail to re-emerge. Repeated stimulations result in a loss of response due to a lack of reinforcement because calcium ion channels become less responsive. **(4)**

131. Brain development

1 (a) They both act as controls for the studies in Zambia. The black Americans act as a control for the white Americans as well. **(2)**
 These results do not support the hypothesis.
 There is little or no difference between the scores of the black and white children from the USA. **(2)**

132. Brain chemicals

1 (a) L-Dopa can reach the brain unlike dopamine. In the brain it is converted to dopamine. It reduces symptoms of Parkinson's disease, which is characterised by low dopamine levels. **(2)**
 (b) MDMA gives higher levels of serotonin because it inhibits reabsorption of serotonin into the neurone and may cause more serotonin to be released. **(3)**
 (c) (i) The treatment with the drug mimics Parkinson's disease symptoms. **(1)**
  (ii) The rationalist view is that any overall good must outweigh any overall harm, in this case, harm to the animals. The absolutist view would suggest that any use of this is unacceptable. **(2)**
 (d) First of all, a small sample should be tested to check for safety. Large samples are then tested to get information about the effectiveness. Clinical trials on sample sizes in the thousands would now take place using techniques including double blind trials and the use of placebos. Samples must be representative by taking into account such things as age and gender. **(4)**

133. HGP – The Human Genome Project

1 All DNA that would be found in all human beings **(2)**
2 (a) It might prompt her to more regular self-examinations or attendance for medical checks. She may also be prompted to avoid high-risk practices such as drinking alcohol, hormone replacement therapy and smoking. It may also encourage her to take plenty of exercise, particularly after the menopause. **(3)**
 (b) The patient may feel strongly about data protection and be afraid that third parties, such as insurance companies, may have access to this information. This knowledge may cause anxiety when in fact the information is about risk and not certainty. **(3)**

3 Differences in the way people respond to different drug treatments may be due to underlying differences in their genomes. In some the efficacy may be high but there are sever side effects, in others the reverse may be true, or there could be other combinations of efficacy and side effects. If a patients genome is known, it may be possible for the doctor to select the best drug for that patient, thus personalising their treatment. Another possibility is that, with a knowledge of susceptibilities, doctors may be able to give specific life-style advice.

134. Genetically modified organisms

1 (a) Genetically modified cotton is likely to have an improved yield, which is an advantage to the farmer. It may also allow the farmer to spend less money on insecticides, further increasing his profit. The reduced use of insecticides would have benefits for wildlife as it would not kill insects that are not pests, for example, bees. **(3)**

(b) In both, plants with desired characteristics are selected and this is done deliberately to improve yield. Conventional plant breeding is done by transferring pollen between two plants that have the desired characteristics. Genetic engineering uses techniques such as gene guns and microbial vectors. Genetic engineering is capable of combining genes from a wider range of sources and results are much quicker than with conventional plant breeding. **(4)**

(c) There is little convincing evidence that GM is harmful and it is widely used in the USA and India. GM has huge potential to reduce prices. In addition, it can reduce dependency on pesticides and therefore damage to the environment. It has already given benefits such as the production of human insulin by bacteria.
On the other hand, there could be undiscovered risks of GM and not all GM projects have been successful. Genes may be transferred from engineered plants to other organisms with consequences that we cannot foresee. There is always the potential for exploitation of poorer farmers by biotech companies. There may be a loss of genetic diversity as we rely more and more on a small number of GM strains. **(9)**

135. Exam skills

1 Fewer calcium ions, **4**, will enter the pre-synaptic region through calcium channels, **3**.
This will mean fewer vesicles, **5**, containing neurotransmitter, **6**, will fuse with the pre-synaptic membrane, **1**.
As a result, less neurotransmitter diffuses across the synaptic cleft and so less binds to sodium channels, **7**.
Fewer channels will open and so fewer sodium ions will diffuse through the post-synaptic membrane, **2**.
This will result in less depolarisation and fewer action potentials. **(5)**

2 (a) An action potential arrives at neuromuscular junction. This results in the secretion of a neurotransmitter which stimulates the sarcoplasmic reticulum to release calcium ions.

(b) The calcium ions bind with troponin and cause movement of tropomyosin. This allows cross bridges between actin and myosin to form by exposing the actin.

136. AS Level Timed test 1

1 (a)

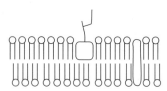

phospholipid bilayer – correct orientation;
glycoprotein – in outer layer only;
intrinsic protein – spanning both layers **(3)**

(b) rate increases when the concentration increases;
this increases the concentration gradient;
levelling off of curve due to channel proteins being saturated with molecules of the substance;
no more can be carried per unit time **(2)**

2 (a) (i) Any three from:
thick wall;
lumen;
endothelium;
smooth muscle;
elastic fibres;
tunica media;
connective tissue;
tunica adventitia **(3)**

(ii) thick wall to withstand blood under high pressure;
narrow lumen to maintain high pressure;
presence of elastic fibres to allow vessel to stretch;
smooth muscle contracts to squeeze blood along;
smooth lining reduces friction;
folded lining allows artery to stretch without tearing **(2)**

(b) reduced blood flow;
less oxygen reaches brain;
less glucose reaches brain;
less aerobic respiration;
less ATP produced;
brain needs lots of energy to function;
lactic acid produced from anaerobic respiration;
lactic acid inhibits enzymes **(3)**

(c) (i) the alleles present in an organism: in a 719 Arg carrier they would be one 719 Arg allele and one normal **(1)**

(ii) a different form of one gene: in 719 Arg there is a normal allele and the 719 Arg allele **(1)**

(d) Two from:
muscle inflammation;
liver damage;
joint aches and pains;
nausea;
constipation;
diarrhoea;
kidney damage;
cataracts;
diabetes;
allergies;
respiratory problems;
headaches;
depression **(2)**

3 (a) (i) make named condition that affects rate non-optimal, e.g. pH or temperature **(1)**

(ii) hydrolysis **(1)**

(iii) sucrose and water;
fructose and glucose **(2)**

4 (a) (i) D **(1)**

(ii) B **(1)**

(iii) A **(1)**

(b) The antisense strand of the DNA which makes up the gene for the protein is transcribed into mRNA.
The mRNA passes into the cytoplasm and joins with a ribosome.
tRNA brings amino acids to the mRNA and complementary bases pairing joins the tRNA with the correct anticodon in position on the codon on the mRNA. The amino acid on the tRNA is joined to the next one in sequence via a peptide bond.
This process continues until a stop codon is reached when the complete polypeptide is released. **(5)**

5 (a) since 32% G then 32% also C so 64% G+C, so 36% A+T so 18% are A
uracil pairs with adenine so mRNA will also have 18% so 18% of 450 = 81 uracil bases **(3)**

(b)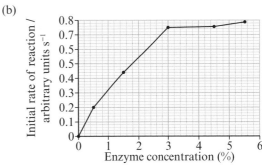

(3)

(c) The two strands of DNA unwind and split apart.
Free nucleotides line up along each strand observing the complementary base pairing rules (Watson and Cricks' specific pairing).
Nucleotides join together as a phosphodiester bond forms between each deoxyribose and adjacent phosphate group.
Hydrogen bonding links the two strands together. **(4)**

6 (a) (i) **B (1)**
   (ii) **D (1)**
   (iii) **B (1)**
   (iv) **C (1)**
  (b) (i) so that only one factor has changed;
      if intake went up this could increase risk **(2)**
     (ii) both diets decrease the risk;
      both diets have less saturated fats;
      saturated fat associated with heart disease;
      changing to unsaturated lipids has the greater effect;
      excess carbohydrates may be stored as saturated lipids;
      unsaturated lipids change HDL/LDL ratio **(3)**

7 (a) Fenitrothion increases the permeability of membranes.
   The cell surface membrane is affected before the vacuole membrane.
   Mineral ions leak out immediately because only the cell surface membrane needs to be damaged.
   For betalain leakage, both the cell surface membrane and the vacuole membrane would need to be damaged so there is a delay. **(4)**
  (b) Indicative content:
   reference to three or more different pH values used as it is the independent variable;
   validity ensured by detailing how pH varied, e.g. buffer;
   repeats at each pH to estimate reliability of data;
   obtain samples from the same beetroot;
   control of other named variable to ensure validity, e.g. temperature, concentration of fenitrothion, size of discs;
   safety aspect explained, e.g. (cut finger when cutting beetroot + cut away from finger/cut on a hard surface) **(6)**

8 (a) bar showing 2%;
   bar showing 16%;
   obesity, dark and overweight, light portions identified **(3)**
  (b) (i) **A (1)**
     (ii) **USA (1)**
     (iii) **Australia (1)**
     (iv) **18:17 (1)**
  (c) graph shows percentages;
   population size not known;
   may be a different number of males to females **(2)**
  (d) (i) The relationship between two variables is such that a change in one of the variables is reflected by a change in the other variable. **(1)**
     (ii) The consumption of corn syrup goes up.
      This is before the increase in obesity.
      The consumption of dextrose falls with time, e.g. during the 1970s.
      The consumption of glucose stays fairly constant. **(3)**

9 (a) gradient = rate = 40 arbitrary units ÷ 90 s
   = 0.44 arbitrary units s^{-1}
   enzyme concentration this graph represents is 1.5% **(3)**

axes correct (x = enzyme concentration, y = initial rate of reaction) and labelled with units;
plots all correct;
suitable line drawn **(3)**

(c) as enzyme concentration rises there are more active sites;
therefore, more substrate can be converted per unit time;
at about 3.0% substrate becomes limiting so rate does not rise any further **(3)**

AS Level Timed test 2

1 (a) (i) genetic differences;
      molecular differences;
      large number of differences – Archaea are more different from Bacteria than humans are different from a crab or spider **(3)**
     (ii) Scientific findings:
      published in a journal;
      presented at scientific conference;
      peer reviewed;
      other scientists repeat experiments to validate findings **(3)**
  (b) **A (1)**

2 (a) (i) both carry genetic material as haploid set of chromosomes;
      sperm motile, egg not;
      sperm much smaller than egg **(3)**
     (ii) sperm cell fuses with egg cell membrane;
      cortical granules;
      move towards egg cell surface membrane;
      exocytosis of cortical granules;
      contents of cortical granules secreted into jelly layer;
      called the cortical reaction;
      hardening of zona pellucida to form fertilisation membrane;
      change in charge across egg cell membrane in some species **(5)**
  (b) (i) **C (1)**
     (ii) Totipotent cells can differentiate to become any cell.
      Pluripotent cannot differentiate to become any cells in the body.
      Only totipotent cells can give rise to other totipotent cells.
      Totipotent cells can give rise to an entire human being, pluripotent cells cannot. **(3)**

3 (a)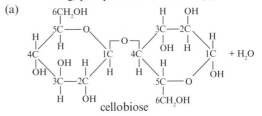

(2)

(b) The monomers are alternately flipped over with respect to the ones they are joined to.
The chains lie next to each other.
Hydrogen bonds hold the chains together to form cellulose microfibrils.
Cellulose has parallel straight chains for a strong structural function in cell walls. **(4)**

(c)

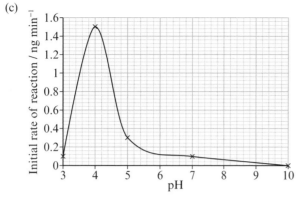

Initial rate calculated as below and plotted:
10 = 0
3 = 0.5 ÷ 5 = 0.1 nanograms min⁻¹
7 = 0.9 ÷ 9 = 0.1 nanograms min⁻¹
5 = 1.2 ÷ 4 = 0.3 nanograms min⁻¹
4 = 3 ÷ 2 = 1.5 nanograms min⁻¹ **(4)**

4 (a) (i)   both have decreased;
             decrease in roadside verges greater than in hedgerows;
             by 0.5;
             percentage decrease greater in roadside verges greater than in hedgerows;
             roadside: (2 ÷ 17) × 100 = 11.8%, hedgerow: (1.5 ÷ 15) × 100 = 10%;
             roadside verges: greater species richness at both beginning and end of decade **(4)**
       (ii) species richness gives no indication of abundance of each species;
             so does not really address biodiversity;
             better to use a calculated biodiversity index;
             number of individuals in each species recorded;
             genetic diversity is also part of biodiversity which is also not addressed in the study **(4)**
   (b) seeds stored in cool, dry conditions for a long time;
        viability tests carried out at regular intervals;
        large numbers of plants can be stored so more economic than conserving   living plants;
        less likely to be damaged by (natural disaster/disease/ herbivores) **(3)**

5 (a) (i)   secretion of waxy substance **(1)**
       (ii) active at night;
             to avoid predation
             OR
             spreads wax over skin;
             to conserve water in a dry habitat
             OR
             hunting in trees rather than on the ground;
             (avoiding high temperatures during the day/finding prey more easily at night) **(2)**
   (b) eats insects at night, in trees;
        within the hot, dry habitat with trees **(2)**
   (c) selection pressures of hot and dry habitat;
        and competition and predation are exerted;
        a mutation in the frog;
        gives rise to an advantageous allele, such as the one for waxy secretions (or other example);
        individuals with the advantageous allele survive and breed;
        and they are passed on to future generations;
        leading to an increased frequency of advantageous alleles in the population **(5)**

6 (a) (i)

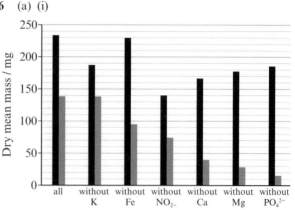

(4)

       (ii) root:shoot ratio with all minerals is 139 ÷ 235 = 0.59;
             all others in this list show a decrease in this value except 'without potassium' (which gives 0.74) – shoot growth is depressed compared to the control, although root growth is not;
             worst is 'without phosphate' at 0.09;
             least affected is 'without nitrate' at 0.53, but both root and shoot growth is depressed compared to the control;
             statement has some truth as it applies to four out of six minerals and these minerals are all needed, so deficiency of any kind is likely to cause plant health issues;
             however, does not taken into account overall size;
             for example, 'without nitrate' shows a high root:shoot ratio of 0.53 but only 202 mg overall mass **(3)**
   (b) calcium used in calcium pectate;
        which is part of middle lamella holding cells together **(2)**

7 (a) **B (1)**
   (b) lactose (B) binds with repressor C;
        C cannot then bind with F, operator gene;
        this allows mRNA polymerase (D) to bind with DNA;
        D then catalyses the transcription;
        of the sense portion of the DNA;
        into the mRNA for β-galactosidase **(6)**

8 (a) (i)   **D (1)**
       (ii) **B (1)**
       (iii) **B (1)**
       (iv) **A (1)**
   (b) (i)   **A (1)**
       (ii) involvement of ribosomes on the rER (X);
             amino acids joined by peptide bonds to form polypeptide chains;
             then folded into 3D shape in rER;
             packaged into vesicles at the end of the rER which move to the Golgi apparatus (Y);
             protein modified in Golgi apparatus;
             modified protein packaged into secretory vesicles (Z) which fuse with cell membrane **(4)**
       (iii) different shape molecule requires active site with different shape;
             cellulose is made of β-glucose and starch is made of α-glucose;
             1,6 glycosidic bonds only in starch;
             starch made of amylose and amylopectin;
             cellulose is linear, starch is branched **(4)**
   (c) cellulose molecules held together by hydrogen bonds in microfibrils;
        microfibrils in parallel;
        a matrix of hemicelluloses and pectin **(2)**

136. A Level Timed test 1

1 (a) genetic material DNA or RNA;
        single-stranded or double-stranded nucleic acid;
        protein coat or capsid/spikes on outside;
        some viruses contain enzymes/envelope may be present **(3)**
   (b) (i)   synthesis of viral components, nucleic acid and proteins;
             assembly of virus;
             both take time **(2)**

(ii) host cell destroyed;
   lots of virus particles are released at the same time;
   more cells can be infected **(2)**

(c) antigen has to attach to B cells;
   T helper cells are needed in activation of B cells;
   T helper cells have to be activated before they can activate
   B cells; cloning of B cells has to take place;
   B cells specialise into plasma cells;
   antibody production/secretion by plasma cells;
   all take time **(4)**

2 (a) (i) rate of production of biomass;
      losses in respiration in producers **(2)**

   (ii) NPP depends on photosynthesis;
      the higher the temperature the more NPP;
      enzymes in photosynthesis can work faster;
      increase in rainfall increases NPP;
      water needed for light-dependent reaction;
      the role of water in transport **(5)**

   (iii) the shape would be similar to NPP graph;
      the line would be higher than NPP graph;
      GPP has to be higher than NPP as respiration has to
      be subtracted from GPP **(3)**

(b) $2800 - 1750 = 1050$
   $(1050 \div 5300) \times 100 = 19.8\%$ **(2)**

3 (a) (i) Chlorophyll is excited by a gain in light energy which
      results in excited electrons.
      They leave the chlorophyll molecule.
      They get picked up by molecule A. **(2)**

   (ii) photolysis using light energy of water;
      hydroxide ions as source of electrons **(2)**

   (iii) As electrons move from carrier to carrier, energy is
      released to synthesise ATP from ADP + Pi.
      H^+ is moved into lumen of thylakoid using light energy.
      H^+ diffuses out of thylakoid synthesising ATP. **(3)**

(b) (i) RuBP required for carbon dioxide fixation;
      catalysed by RuBP carboxylase;
      to provide GP;
      reduced NADP used to reduce GP;
      to GALP which also requires ATP;
      and is used to form glucose;
      GALP also used to regenerate RuBP **(5)**

   (ii) **A (1)**
   (iii) **C (1)**

4 (a) (i) fatty acids;
      glycerol **(1)**

   (ii)
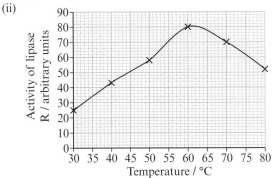

   axes correct;
   plots correct;
   temperature increase causes a rise to an optimum then a fall;
   optimum at 60°C;
   increasing kinetic energy causes the rise;
   enzyme denaturation causes the fall **(5)**

(b) (i) **D (1)**
   (ii) **D (1)**
   (iii) **C (1)**
   (iv) **A (1)**

5 (a) **B (1)**
(b) **D (1)**

(c) Indicative content:
   lichens and mosses as pioneer community;
   (able to) grow in (little/no) soil/equivalent;
   (that) breaks up (rock) fragments/forms (thin/shallow/
   equivalent) soil;
   (plants/equivalent) with (small/short/equivalent) roots;
   (able to) grow in (thin/shallow/equivalent) soil/equivalent;
   changes in soil structure enable (trees/shrubs) to grow/
   equivalent;
   soil able to (hold/retain/contain/equivalent) (water/minerals);
   as plants (lose leaves/die/decay/equivalent);
   (organic matter/humus/equivalent) (increases/releases/
   equivalent);
   competition effects **(6)**

(d) includes (both) animals and plants/has many species/has
   high biodiversity/equivalent;
   (interaction/equivalent) between species/equivalent;
   (dominant/codominant) (plant or animal) species;
   stable if no (change to environment/human influence) **(4)**

6 (a) the role of the sea anemone;
   it is a predator;
   it might control population of prey;
   it is also prey and provides food for other animals/provides
   shelter for some animals **(3)**

(b) reduces surface area to volume ratio so less water loss;
   also reduces visibility to predators;
   no need for tentacles to be exposed so energy will be conserved
   (3)

(c) (i) abiotic **(1)**
   (ii) no indication that temperature has an effect;
      as shows little variation, only 2°C;
      distribution is correlated with height above sea level;
      more likely to dry out at higher levels;
      food availability differs, e.g. less at higher levels, more
      at lower levels;
      more likely to be eaten at lower levels **(3)**

7 (a) In 1970s, red areas disappear.
   In 1980s, red areas increase towards the end.
   In 1990s, red areas increase to 1995 then decrease. **(3)**

(b) Temperature affects enzyme activity.
   Therefore, the life cycle of the beetles is affected.
   The availability of food varies with temperature. **(2)**

(c) Before 1970, the temperature was mostly below the mean
   and there was no 'red area'.
   Before 1970, the drought index was mostly below the mean
   and there was no 'red area'.
   As the temperature increases so does the 'red area'.
   As the drought index increases so does the 'red area'. **(3)**

(d) relatively short period of time;
   data only relates to Alaska;
   beetles may be affected by another factor;
   only a correlation;
   periods of drought and high temperatures do not always
   coincide with years with large areas of 'red area';
   fluctuations in data;
   no information about number of measurements of
   temperature and drought index **(3)**

8 (a) use of quadrats;
   method for placing quadrats randomly described, such as
   use of random co-ordinates in a grid;
   method to record percentage cover described, such as
   gridded quadrat;
   estimate made over time **(4)**

(b) (i) change in percentage cover from graph is −60%
      60% of 24% is 14.4 percent cover after ten years
      so 14 ÷ 10 5 per year = 1.4% year⁻¹ **(3)**

   (ii) t-test;
      because t-test is the test for the significance of the
      difference between two samples **(2)**

(c) will probably decrease;
   because only 10% up with water but 40% down with heat **(2)**

9 (a) The membrane is fluid.
  The fluidity allows the membrane to change shape.
  Fluidity allows membranes to fuse.
  Proteins in the membrane play a role in cell transport. **(4)**
  (b) use a microscope to count the number of yeast cells at start and end of investigation;
  use a range of five temperatures in incubators;
  yeast are left for a period of time for budding to occur;
  repeat to calculate a mean number of yeast cells;
  describe calculation of rate of asexual reproduction;
  name control variable **(5)**

A Level Timed test 2

1 (a) (i) **B (1)**
    (ii) **D (1)**
    (iii) **A (1)**
    (iv) **D (1)**
    (v)   **A (1)**
  (b) muscles work antagonistically;
  circular muscle relaxes;
  radial muscle contracts **(2)**

2 (a) enzyme 1 converts P to Q OR enzyme 2 converts Q to R OR enzyme 3 converts R to S which becomes the next substrate;
  enzyme shows specificity;
  enzyme controls the conversion;
  enzyme speeds up the conversion;
  by reducing activation energy **(4)**
  (b) (i) W is NADox;
    formed due to reduced NAD releasing electrons;
    which go to carrier A;
    H^+ is moved into inter-membranal space **(3)**
    (ii) H^+ pass through stalked particle which is ATP synthase;
    H^+ passes down an electrochemical gradient;
    sufficient energy is released;
    to join ADP and Pi **(3)**
  (c) (i)   **C (1)**
    (ii) **A (1)**
    (iii) **C (1)**
    (iv) $48 \div 60$
      $= 0.8$ mm per minute **(2)**
    (v)   masses of organisms may differ so use same mass;
      temperature may change so control temperature in a thermostatically-controlled water bath/equivalent;
      pressure may affect volume of gas so use a control with no organisms at the same time **(3)**

3 (a) (i)   **D (1)**
    (ii) **C (1)**
    (iii) medulla **(1)**

4 (a) (i)   **B (1)**
    (ii) **D (1)**
    (iii) **A (1)**
    (iv) It is myelinated with nodes so can show saltatory conduction in which the impulse jumps from node to node.
      This increases the speed of the impulse. **(2)**
  (b) (i) action potentials not possible;
    sodium ions move into axon;
    down a concentration gradient;
    neurone is permanently depolarised;
    resting potential cannot be reestablished **(4)**
    (ii) the frogs' ion channel protein is different;
    so the poison cannot bind;
    maybe the poison is broken down **(2)**

5 (a) (i) at 120 sec it is 0.8 dm^3, at 0 sec it is 1.6 dm^3
    $1.6 - 0.8 = 0.8$ dm^3 in 120 sec
    so consumption is $0.8 \div 2 = 0.40$ to 0.42 dm^3 (min^{-1}) **(3)**
    (ii) $(0.4 \div 70) \times 60$
    $= 0.34$ dm^3 kg^{-1} h^{-1} **(3)**
    (iii) It has:
    more peaks closer together;
    higher peaks;

a steeper slope **(3)**
  (b) (i) soda lime **(1)**
    (ii) there would be no downward slope;
    because exhaled carbon dioxide equals consumed oxygen
    OR
    tidal volumes increase;
    due to increase in carbon dioxide concentration **(2)**
  (c) carbon dioxide increase in blood leads to fall in blood pH;
  detected by chemoreceptors in the medulla ventilation centre;
  from which impulses are sent;
  to intercostal muscles and diaphragm;
  leads to increased breathing rate and depth **(5)**

6 (a) (i) Slow twitch has (any 3 from):
    more mitochondria;
    more myoglobin;
    less sarcoplasmic reticulum;
    more capillaries;
    less glycogen;
    less creatine phosphate;
    more resistant to fatigue **(3)**
    (ii) prey **(1)**
    (iii) predators have more fast twitch than slow twitch;
    anaerobic respiration;
    glycolysis is used **(3)**
  (b) lactate is removed from muscle and converted to pyruvate;
  using NADH to produce carbon dioxide and water;
  and enters Krebs cycle;
  some lactate is converted to glycogen in the liver **(3)**
  (c) changes to core temperature are detected by the hypothalamus thermoregulatory centre;
  arterioles vasodilate;
  which increase blood flow to the skin;
  this increases loss of heat from skin;
  which has a cooling effect;
  this is negative feedback **(4)**

7 (a) columns are (smaller/narrower/equivalent) (in visual cortex for left eye);
  (sensory) (neurone/axon) (shorter neurone/reduced growth) (for left/deprived eye);
  (fewer/shorter dendrites)/fewer (synapses/branches) (in left/deprived eye)/equivalent **(3)**
  (b) (critical/sensitive) period/critical window (in visual development);
  if one eye is deprived of (stimulation/light), (neurones/dendrites/synapses/columns) do not develop/equivalent **(2)**
  (c) visual deprivation studies;
  e.g. cataract removal from children, bandaging of eyes, reference to development of distance perception,
  e.g. Müller-Lyer **(2)**

8 (a) (i) single agar block with no IAA **(1)**
    (ii) Indicative content:
    both sides of shoot taller;
    than the control;
    both IAAs diffuse down;
    causing cell elongation;
    by water uptake;
    the shoot bends to the right;
    due to more growth on left side of shoot **(6)**
  (b) IAA enters the cell;
  binds to transcription factor and forms a transcription initiation complex;
  switching on gene;
  by promoter region;
  which allows formation of mRNA translation, which produces protein **(4)**

9 (a) both show decrease in time siphon withdrawn with repeated trials;
  sea slug from rough water withdrawal time is lower;
  decrease in time siphon withdrawn is faster in sea slug from calm water;
  this is called habituation **(4)**

(b) stimulus in rough water is harmless;
less withdrawal of siphon saves energy;
and allows gas exchange **(3)**
(c) fMRI involves brain activity in real time.
fMRI measures oxygen uptake.
The active area of brain gets more blood.
Oxyhaemoglobin and deoxyhaemoglobin are involved.
fMRI uses radio waves.
The active brain emits less energy.
The more active area appears lighter.
Brain activity falls with habituation. **(5)**

A Level Timed test 3

1 (a) (i) They both act as controls for the studies in Zambia.
The black Americans act as a control for the white
Americans as well. **(2)**
(ii) These results do not support the hypothesis.
There is little or no difference between the scores of
the black and white children from the USA. **(2)**
(b) random/systematic sample of 50 participants from children
living in the rural location of Zambia;
record for every participant of their age, gender, family
occupation;
pair participants on these variables;
every participant tested on the illusion at the start of the
investigation;
one group exposed to a carpentered environment for half
a day every week for eight weeks and given play time in a
formal classroom type setting;
after eight weeks' exposure, participants retested on
the illusion and the differences in the scores before the
investigation and after eight weeks are compared;
compare the change in the control member of the pair with
the experimental (exposed) member of the pair **(5)**

2 (a)

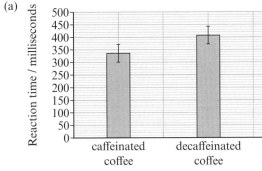

any correct indication range as 246–450 caffeinated,
264–510 decaffeinated **(4)**
(b) because comparing two means;
because the data is normally distributed;
that is, the mean, mode and median will be the same (or
about the same in a sample) **(3)**
(c) degrees of freedom = $(15 + 12) - 2 = 25$
critical value of t with this many degrees of freedom is 2.06
so the null hypothesis should be rejected **(3)**

3 (a) (i) The time taken for the gill to be exposed again is
unaffected by the number of touches. **(1)**
(ii) recovery time falls steeply initially;
stays low **(2)**
(iii) $\Sigma d^2 = 271$, so $6 \times 271 = 1626$
$1626 \div 990 = 1.64$
$1 - 1.64 = -0.64 = r_s$ **(3)**
(iv) for 10 pairs critical value is 0.65
calculated value is -0.64 (sign can be ignored)
so the null hypothesis is accepted **(3)**
(b) ability to withdraw the gill when touched is behavioural;
a type of innate behaviour programmed into the slugs'
genetics;
learning to not withdraw the gill is also behavioural;
a type of learned behaviour;
possession of a gland that releases purple fluid is anatomical
(5)

4 (a) (i) Would need to use:
same masses of senna;
same part/age of senna;
same volume of solvent and water;
same extraction technique/time **(3)**
(ii) all experiments carried out at same temperature;
using an incubator;
all experiments carried out at same pH;
using a buffer **(4)**
(iii) in A it is 12 mm diameter, thus 6 mm radius, area is
therefore 113.1 mm^2
in water it is 2 mm diameter, so radius is 1 mm, area is
3.1 mm^2
so solvent A gave 113.1 − 3.1 mm^2 = 110 mm^2 more
so percentage = $(110 \div 20) \times 100\% = 550\%$ **(3)**
(iv) the diameter cleared by A is about twice as great as
that of C
the SD for A is 0.5 so 2 × SD is 1 and thus the
likelihood of A being less than 12 − 1 = 11 is only 5%
SD for C is 0.2 so 2 × SD is 0.4 so the likelihood of C
being more than 6 is only 5%
so there is no overlap between 2 × SD so we can be
95% sure that the difference is real **(3)**
(b) extract made from Chinese senna in solvent A;
agar plate with bacteria;
agar made up of a range of at least five pH values, two
below and two above 7;
description of aseptic technique;
extract placed on paper disc;
incubated at temperature in range 20 to 30°C for stated
time in range 1 to 7 days;
zone of inhibition;
replication at each pH to measure variability **(5)**

5 (a) fatty acid tails are hydrophobic;
so orientate themselves away from water which is a polar
environment;
phosphate heads are hydrophilic;
so can interact with water;
cytoplasm and tissue fluid are watery and form a polar
environment **(3)**
(b) monolayer film of phospholipids is twice as large as the cell
surface area;
supporting the idea of a bilayer;
microscope images of cell surfaces show proteins sticking out;
supporting the idea of proteins floating in the lipid layer;
when lectins, which react with carbohydrates, are added to
a membrane they are found only on the outside;
supporting the idea of glycoproteins on the outside only;
some water-soluble substances pass into and out of cells;
supporting the idea of hydrophilic channels **(4)**
(c) Indicative points:
provides boundary;
carrier in membrane able to select ions or molecules and
transport through membrane;
protein channels as selective pathways through membranes;
enzymes embedded in the membrane;
e.g. digestive enzymes, such as maltase, in the cells lining the
small intestine or ATPase in the cristae of the mitochondria;
compartmentalisation;
internal membranes surround organelles;
organelles example may have internal membranes;
localisation of enzymes;
e.g. lysosomes, acrosome;
the need to isolate enzymes from rest of cytoplasm to
prevent autodigestion;
having enzymes confined to organelles **(9)**

6 (a) They have large incisor teeth to tackle fibrous plant
material and to dig. This is an anatomical adaptation.
They can tolerate low levels of oxygen which is physiological.
Some individuals, called soldiers, defend the colony against
predators whilst others find food, which is behavioural. **(3)**

(b) the role of the naked mole rat within the ecosystem;
this includes its feeding role, shown by the trophic level it occupies;
which is primary consumer;
it is also prey and provides food for other animals;
its burrows may provide shelter for other animals **(3)**

(c) it shows that the information in that sentence has come from a scientific journal;
which will have been peer reviewed **(2)**

(d) Oxygen is used as the final electron carrier in the electron transport chain.
Without oxygen, the electron transport chain will stop functioning.
This means that NADoxidised cannot be regenerated.
Without NADoxidised the Krebs cycle cannot function and so aerobic respiration stops.
Anaerobic respiration can still happen but the production of ATP is much lower than in aerobic.
ATP is the form in which energy is transferred in cells.
So, energy-requiring processes would stop or be reduced if ATP concentration was lowered or none was produced. **(5)**

(e) selection pressure is an environment with very low oxygen levels;
mole rats show genetic variability in their ability to withstand low oxygen levels;
due to mutations;
mole rats which were able to withstand very low oxygen levels were more likely to survive and reproduce;
this means that alleles for withstanding oxygen deprivation would increase in the population **(4)**

(f) the queen is within the environment of non-breeders;
an interaction between them leads to suppression of gonadotrophin-releasing hormone;
probably by a gene switching mechanism;
this in turn suppresses ovulation and sperm production in the phenotype;
if non-breeders are isolated in an environment of mixed-sex pairs then the phenotype will change to fertile individuals;
again as a result of gene switching **(4)**

(g) the signals may turn on the early contact inhibition gene p16;
they act as transcription factors which;
interact with RNA polymerase to switch the gene on;
they may do this by entering the cell;
or through a second messenger **(5)**

(h) Inbreeding is reproduction between individuals who have a very similar genotype, T.
It leads to increased homozygosity.
This may lead to an increase in the chances of offspring having recessive alleles in pairs which may be deleterious.
This leads to a decreased fitness of a population (called inbreeding depression). **(4)**

(i) Utilitarianism:
suffering is restricted to few individuals;
mole rats have reduced sensitivity to chemical pain therefore have less capacity to suffer;
research on mole rats has very large potential benefits to human health **(3)**

Published by Pearson Education Limited, 80 Strand, London, WC2R 0RL.

www.pearsonschoolsandfecolleges.co.uk

Copies of official specifications for all Edexcel qualifications may be found on the website: www.edexcel.com

Text and illustrations © Pearson Education Limited 2016
Typeset by Kamae Design
Produced by Out of House Publishing
Illustrated by Tech-Set Ltd, Gateshead
Cover illustration by Miriam Sturdee

The rights of Gary Skinner and Ann Skinner to be identified as authors of this work have been asserted by them in accordance with the Copyright, Designs and Patents Act 1988.

First published 2016

19 18 17 16

10 9 8 7 6 5 4 3 2 1

British Library Cataloguing in Publication Data
A catalogue record for this book is available from the British Library

ISBN 978 1 4479 9270 7

Printed in Slovakia by Neografia

Acknowledgements
The publisher would like to thank Edexcel for permission to reproduce extracts from the following Salters-Nuffield AS exam papers: Jan 2004; Jan 2009; May 2011; Jan 2010; Jan 2012; May 2012; May 2013; Jan 2011; June 2015; June 2013; June 2014; June 2013; June 2009; Jan 2014; Jan 2006; Jan 2004; June 2011; Jan 2008; June 2004; June 2007. Salters-Nuffield A papers: June 2007; June 2008; June 2005; Jan 2007; June 2015; Jan 2010; Jan 2012; Jan 2013; June 2014; June 2006; June 2011; June 2006; June 2010; Jan 2008; Jan 2015; Jan 2005; June 2012; Jan 2014; June 2005; Jan 2015; June 2014; June 2010; May 2012; Jan 2011; June 2009.

We are grateful to the following for permission to reproduce copyright material:
Figure in Topic 1 on p. 9 The effect of having a high BMI on the risk of death due to CHD compared with a control group with a BMI of 20, graph, http://apps.who.int/gho/data/node.home World Health Organisation; figure in Topic 4 on p.7 Sparing and sharing forest management strategies, graph, http://www.cell.com/cms/attachment/2017446092/2037759148/gr1b3.jpg, Cell Press (Elsevier).

The publisher would like to thank the following for their kind permission to reproduce their photographs:

(Key: b-bottom; c-centre; l-left; r-right; t-top)

Ardea: David Northcott / Danita Delimon 145; **Gary Skinner**: 64; **Getty Images**: George Chapman 82; **Rex Shutterstock**: Everett Collection 73l; **Science Photo Library Ltd**: Biology Pics 19, Biophoto Associates 148, CNRI 2, Don Fawcett 44, Dr Juerg Alean 73r, Dr. John D. Cunningham, Visuals Unlimited 155, Herve Conge, ISM 49, Living Art Enterprises 128; **Shutterstock.com**: Piotr Krzeslak 120

All other images © Pearson Education

A note from the publisher
In order to ensure that this resource offers high-quality support for the associated Pearson qualification, it has been through a review process by the awarding body. This process confirms that this resource fully covers the teaching and learning content of the specification or part of a specification at which it is aimed. It also confirms that it demonstrates an appropriate balance between the development of subject skills, knowledge and understanding, in addition to preparation for assessment.

Endorsement does not cover any guidance on assessment activities or processes (e.g. practice questions or advice on how to answer assessment questions), included in the resource nor does it prescribe any particular approach to the teaching or delivery of a related course.

While the publishers have made every attempt to ensure that advice on the qualification and its assessment is accurate, the official specification and associated assessment guidance materials are the only authoritative source of information and should always be referred to for definitive guidance.

Pearson examiners have not contributed to any sections in this resource relevant to examination papers for which they have responsibility.

Examiners will not use endorsed resources as a source of material for any assessment set by Pearson.

Endorsement of a resource does not mean that the resource is required to achieve this Pearson qualification, nor does it mean that it is the only suitable material available to support the qualification, and any resource lists produced by the awarding body shall include this and other appropriate resources.

For your own notes